51 Structure and Bonding

Electron Nuclear Double Resonance of Transition Metal Complexes with Organic Ligands

By A. Schweiger

With 47 Figures and 19 Tables

Springer-Verlag
Berlin Heidelberg New York 1982

ISBN 3-540-11072-0 Springer-Verlag Berlin Heidelberg New York
ISBN 0-387-11072-0 Springer Verlag New York Heidelberg Berlin

Library of Congress Catalog Card Number 67-11280

Typesetting and printing: Schwetzinger Verlagsdruckerei. Bookbinding: J. Schäffer OHG.
2152/3140-543210

Table of Contents

Electron Nuclear Double Resonance of Transition Metal Complexes with Organic Ligands

Arthur Schweiger

Laboratory for Physical Chemistry, Swiss Federal Institute of Technology,
CH-8092 Zürich, Switzerland

Electron nuclear double resonance (ENDOR) of transition metal complexes with organic ligands is considered from an experimental and theoretical point of view. In the experimental sections, instrumentations used in conventional ENDOR and multiple resonance spectroscopy and in some new double resonance methods are discussed. Particular emphasis is placed on various experimental techniques which supplement the conventional ENDOR experiment. The different methods are illustrated by a number of applications on transition metal complexes[1]. In the theoretical sections, the analysis of ENDOR spectra of complex spin systems is described including second order effects, transitions probabilities and determination of signs of the magnetic parameters. Some of the approaches used in ENDOR to interpret ligand hyperfine and quadrupole data are outlined.

Finally, a review of the ENDOR literature on transition metal complexes with organic ligands is presented.

1 ENDOR represents an alternative and very sensitive method to study quadrupole couplings of ligands and central ions in paramagnetic transition metal complexes

1 Introduction

Since the first observation of ligand hyperfine (hf) interactions in the EPR[2] spectrum of a transition metal compound by Owen and Stevens in 1953[1)], a great amount of EPR work has been reported on this subject. In the last three decades this method has proved to be particularly sensitive for probing the details of the environment of paramagnetic transition metal ions[2)]. From the electron **g** tensors, the zero-field splittings and the hf interactions of central ions and ligands, precise information about the geometric and electronic structure could be evaluated for a large variety of compounds.

One disadvantage of the EPR spectroscopy lies in the fact that in rigid media the EPR lines are inhomogeneously broadened by the large number of nuclei which interact with the unpaired electron(s). As a consequence, ligand hf couplings are often not resolved in the EPR display. In addition, the specific selection rules which characterize the EPR spectrum generally prevent the detection of nuclear quadrupole interactions[2)].

To resolve hf and nuclear quadrupole interactions which are not accessible in the EPR spectra, George Feher introduced in 1956 a double resonance technique, in which the spin system is simultaneously irradiated by a microwave (MW) and a radio frequency (rf) field[3)]. This *e*lectron *n*uclear *do*uble *r*esonance (ENDOR) spectroscopy has widely been applied in physics, chemistry and biology during the last 25 years. Several monographs[2, 4–6)] and review articles[7–11)] dealing with experimental and theoretical aspects of ENDOR have been published.

The principle of the ENDOR method is illustrated in Fig. 1. It refers to the most simple spin system with an electron spin $S = 1/2$ and a nuclear spin $I = 1/2$ for which an isotropic hf interaction, a_{iso}, is considered. In a steady state ENDOR experiment[4)], an EPR transition (A, D), called the *observer,* is partly saturated by microwave radiation of amplitude B_1 while a driving rf field of amplitude B_2, called the *pump,* induces nuclear transitions. At frequencies ν_1 and ν_2, the rf field tends to equalize the populations within the m_s-states. This alters the degree of saturation of the observer so that, in the display of the EPR signal height versus the radio frequency, two ENDOR lines at transition frequencies $\nu_1 = |a_{iso}/2 - \nu_n|$ (A, B) and $\nu_2 = |a_{iso}/2 + \nu_n|$ (C, D) will be observed ($\nu_n = \mu_N g_n B_0$ denotes the nuclear Zeeman frequency for a static field B_0).

One requirement for successful ENDOR is the partial saturation of both the EPR and the NMR transitions defined by

$$\gamma_e^2 B_1^2 T_{1e} T_{2e} \geq 1 \tag{1.1}$$

and

---

2 Abbreviations used throughout the text will be given in Appendix A

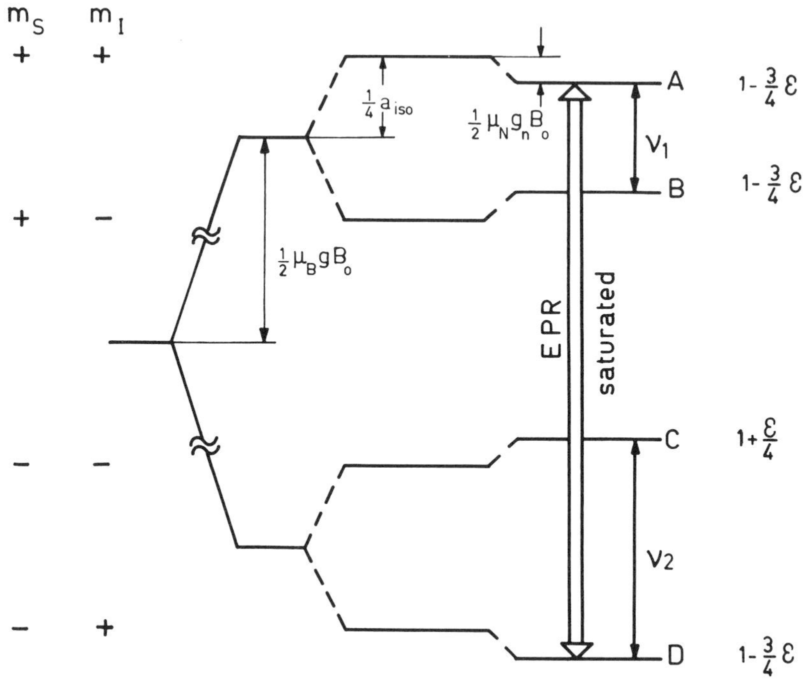

Fig. 1. Energy level diagram for an S = 1/2, I = 1/2 spin system with an isotropic hfs $a_{iso} > 2\mu_N g_n B_0 > 0$. The steady state populations shown are for the case where the (*A, D*) (EPR observer) and the (*A, B*) (rf pump) transitions are saturated

$$\gamma_n^2 B_2^2 T_{1n} T_{2n} \geqslant 1 \ , \tag{1.2}$$

respectively. These conditions can often be fulfilled by appropriately chosen experimental parameters. In transition metal complexes, the spin-lattice relaxation times, T_{1n} and T_{1e}, are usually too short at room temperature to fulfill (Eq. 1.1) and (Eq. 1.2) so that ENDOR experiments for this type of compounds have to be performed almost exclusively at temperatures below 30 K. Since the ENDOR signal is only a small fraction, Δ (ENDOR enhancement), of the EPR signal intensity (typically 1% for transition metal complexes) a further requirement for ENDOR is that the EPR signal-to-noise ratio should be greater than 1/Δ. In this work we shall only be concerned with ENDOR frequencies and *relative* ENDOR intensities, so that a detailed theory of the ENDOR enhancement will be omitted. For spin dynamics associated with ENDOR observations, the reader should consult the literature on the subject[4, 5].

For more complicated spin systems than the one presented in Fig. 1, the number of EPR lines becomes much greater than the number of ENDOR lines. This is due to the different selection rules for EPR and ENDOR transitions. In an EPR spectrum ($\Delta m_S = \pm 1$, $\Delta m_I = 0$) the number of lines increases multiplicatively with the number of nonequivalent nuclei, whereas in ENDOR ($\Delta m_S = 0$, $\Delta m_I = \pm 1$) the corresponding increase is only additive[7]. Since on the other hand the total spectral widths in EPR and ENDOR are of the same order of magnitude (expressed in units of energy), the average spectral density is much smaller in the ENDOR than in the EPR display. Thus, hf couplings which can *not* be observed in a single crystal EPR spectrum, because of inhomogeneous line broadening, may very frequently be resolved with ENDOR.

For compounds containing nuclei with $I \geq 1$, the ENDOR spectrum is characterized by *first order splittings* due to nuclear quadrupole interactions. In EPR this coupling gives rise to small and usually not detectable second order shifts. Information on nuclear quadrupole interactions is normally obtained from NQR. In paramagnetic compounds, however, this spectroscopy suffers from the short relaxation times and from poor sensitivity (considerably lower than in ENDOR). Thus, ENDOR represents an alternative and very sensitive method to study quadrupole couplings of ligands and central ions in paramagnetic transition metal complexes.

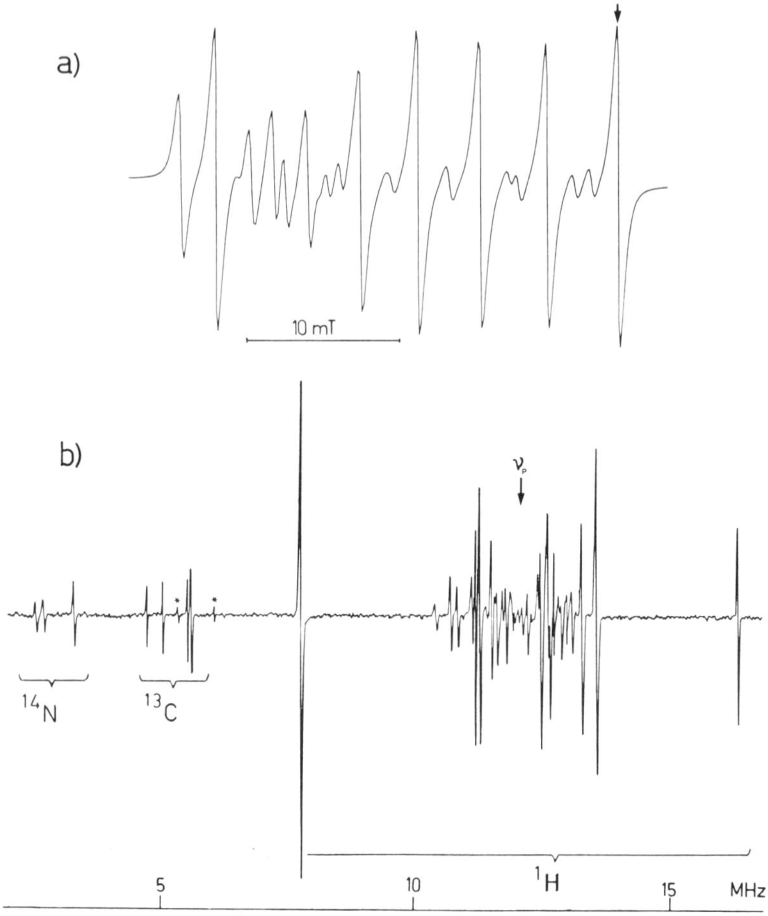

Fig. 2a, b. EPR and ENDOR spectrum of the low-spin Co(II) *Schiff* base complex Co(acacen) diluted into a Ni(acacen) · $1/2\,H_2O$ single crystal, temperature 8 K. **a)** EPR spectrum; the two magnetically nonequivalent sites coincide for this particular orientation (EPR observer is marked by an arrow); **b)** ENDOR spectrum of ^{1}H, ^{13}C (enriched) and ^{14}N ligand nuclei; ν_p: free proton frequency; * denote the $\Delta m_N = \pm 2$ nitrogen ENDOR transitions. (From Ref. 12)

A typical example of resolution enhancement by ENDOR is given in Fig. 2. It compares the EPR spectrum of the cobalt(II) low-spin *Schiff* base complex Co(acacen) with its corresponding ENDOR spectrum[12]. In the EPR spectrum (Fig. 2a) the observed splittings are due to magnetic interactions of the cobalt nucleus ($I_{Co} = 7/2$, "allowed" $\Delta m_{Co} = 0$ and "forbidden" $\Delta m_{Co} = \pm 1, 2$ transitions). No ligand hf couplings, either from nitrogens directly coordinated to the Co(II) ion or from the surrounding protons, are resolved.

In the ENDOR spectrum (Fig. 2b), however, the resolution is greatly improved so that transitions of the two nitrogens, of four carbons (^{13}C enriched), and of a large number of protons with linewidths between 10 and 50 kHz can be detected.

This example impressively demonstrates the great superiority of ENDOR compared to EPR in providing detailed information about the distribution of *spin densities* (hf couplings) and about local *charge densities* (quadrupole couplings) in extended ligand systems. Moreover, ENDOR yields additional information, such as *identification of nuclei* (from the nuclear Zeeman frequency) and evaluation of *relative signs* of hf and quadrupole splittings. The range of applications of the ENDOR technique on transition metal complexes will be elucidated in detail in this monograph.

The first ENDOR spectrum of a transition metal compound was published in 1963 by Deal, Ingram and Srinivasan[13], who studied Cu-phthalocyanine diluted into a single crystal of the corresponding zinc compound. In 1967 Eisenberger and Pershan[14] published some preliminary results on myoglobin azide crystals, but no detailed study of the angular dependence of the ENDOR spectra has been reported. The first complete single crystal ENDOR analysis of a transition metal complex has been presented by Rist and Hyde[15] in 1969, who studied the magnetic interactions of nitrogen and proton ligand nuclei in the planar complex $Cu(Ox)_2$. One year later Hyde's group introduced the idea of orientation selection in ENDOR spectroscopy[16] and published the first ENDOR spectra of a copper protein[17]. Since that time, ENDOR on transition metal complexes has been receiving great attention in inorganic and bioinorganic chemistry. Much work has been done on copper(II) complexes, comprising of Cu(II)-doped ferroelectric crystals and amino acid crystals, complexes with *Schiff* bases and sulfur containing ligands, Cu(II)-porphyrins and corrins as well as copper proteins. Several papers deal with heme compounds and iron-sulfur proteins. Recently, ENDOR studies have been extended to cobalt low-spin *Schiff* base complexes, vitamin B_{12r}, vanadyl complexes and vanadium and titanium sandwich compounds. Most of these compounds have been studied in single crystals, powders or frozen solutions. Recently, the first ENDOR results were reported on transition metal compounds oriented in nematic glasses and on a metal complex in liquid solution.

During the last few years the versatility of ENDOR spectroscopy has been improved by a number of new techniques which make use either of special types of pumping fields (CP-ENDOR, PM-ENDOR), of more than one rf field (DOUBLE ENDOR, multiple quantum transitions, nuclear spin decoupling) or a different display of the spectrum (EI-EPR). In addition to these techniques, alternative methods have been developed (electron spin echo and electron spin echo ENDOR) which are able to supplement or to replace the ENDOR experiment under certain conditions. The utility of all these various advanced techniques, particularly in studies of transition metal compounds, has recently been demonstrated.

The present monograph will be arranged as follows. In Sect. 2 an outline of the instrumentation will be given, specifically for ENDOR experiments in the solid state. Section 3 deals with the analysis of ENDOR spectra based on a perturbation approach. Higher order contributions to ENDOR frequencies and intensities as well as the determination of the signs of the coupling parameters will be discussed. In Sect. 4 we will describe several special ENDOR techniques and illustrate them by examples. The interpretation of hf and quadrupole coupling constants will be outlined in Sect. 5. Finally, in Sect. 6 the ENDOR literature concerning transition metal complexes with organic ligands will be reviewed.

2 ENDOR-Instrumentation

The experimental methods in ENDOR spectroscopy have been extensively described by Kevan and Kispert[4] in their monograph, *Electron spin double resonance spectroscopy*, and by Leniart[18] in a recent paper. In this section we shall briefly review the instrumentation used in solid state ENDOR and describe the technical details of some new experimental methods.

A large number of X-band ENDOR spectrometers have been discussed in the literature. Instruments operating in the less common K-[19], Q-[20, 21] and V-[22] frequency bands have also been described. Most of the ENDOR spectrometers are so-called low-power systems which produce rf fields ≤ 0.1 mT_{rot} (rot: rotating frame). These field strengths are often sufficient to achieve nuclear saturation in transition metal complexes, i.e. to meet the condition (1.2).

For nuclei with a small gyromagnetic ratio γ_n (e.g. ^{14}N, ^{13}C, ^{2}D), the condition (1.2) may be hard to achieve. If this type of nuclei is *strongly* coupled to the unpaired electron(s), however, the applied oscillating rf field, B_2, is *enhanced* by the presence of the electron spin (Sect. 3.3). For small hf couplings, ENDOR intensities may also be increased by lowering the temperature, which corresponds to an increase in T_{ln}. High-power ENDOR spectrometers[23–30] which produce rf fields up to 2 mT_{rot} were originally developed so study liquid phase systems[28]. The commercially available instruments from Bruker[24], Jeol[29] and Varian Instrument Division[30] are of this type. In solid state ENDOR, high rf fields have to be applied in multiple quantum transition and nuclear spin decoupling experiments, as well as in studies of the temperature dependence of ENDOR spectra, e.g. in compounds with intramolecular motions.

2.1 Radio Frequency Range

For ENDOR investigations of ligand and central ion nuclei in transition metal complexes, a large rf range has to be available. For compounds discussed in this work, the transition frequencies of ligand nuclei lie between 1 and 60 MHz. Since hf interactions of central nuclei are often large and strongly anisotropic, their ENDOR transitions may range from the typical ligand ENDOR region mentioned above up to several hundred megacycles.

To cover such a large frequency range, different experimental arrangements are normally used. At frequencies > 50 MHz the mismatch between the power amplifier and the coil arrangement in the cavity often produces serious instability problems due to rf radiation. Consequently a high-frequency broadband ENDOR system has been designed by Berchten[31] to avoid a high voltage standing wave ratio (VSWR) between the power amplifier and the load. The block diagram of this set-up which makes use of a 50–100 Ω

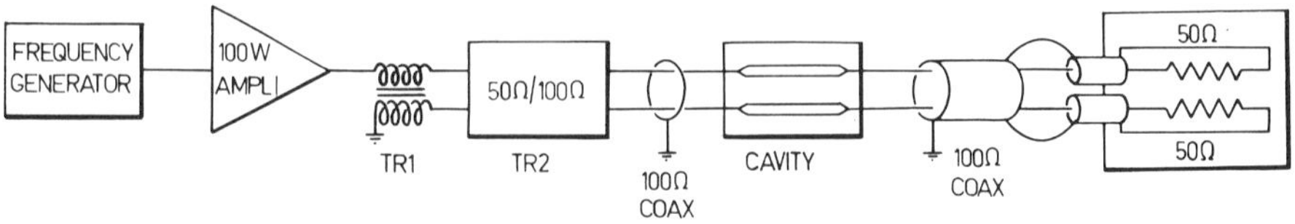

Fig. 3. rf part of the high-frequency ENDOR spectrometer (50–220 MHz). (From Ref. 31)

stepped transmission line transformer (TR 2) with a frequency range 40–220 MHz is shown in Fig. 3. The rf power is fed from this broadband transformer through a symmetrical 100 Ω cable to the coil in a modified E 235 Varian large sample cavity and from the latter to two 50 Ω loads. The four rods in the cavity are connected to form a Lecher type cable with a line impedance $Z_w \simeq 100\ \Omega$. With this arrangement a VSWR of 1/1.2 between the rf transmitter and the loop in the cavity is achieved in the frequency range 70–220 MHz. The rf field at the sample position was measured to be 0.03 mT_{rot}.

2.2 Modulation Schemes

To reduce spurious signals due to drifts of the EPR line setting arising from mechanical and thermal instabilities, double coding of the ENDOR information is often employed[23]. Normally a low-frequency Zeeman modulation (30–300 Hz) is applied while the rf field is frequency or amplitude modulated at frequencies of about 1–30 kHz. This modulation scheme, however, has two major disadvantages:

1) For maximum ENDOR enhancement, the Zeeman modulation amplitude has to be about one half of the width of the EPR line which is saturated at an extremum of its first derivative. However, in an EPR spectrum with line widths of typically 1 mT this Zeeman modulation contributes 20 kHz to the width of a proton ENDOR line. It turns out that in many cases a remarkably better resolution of the spectra may be obtained with a single coding in which only the rf field is modulated.
2) In powder samples with broad EPR lines, large Zeeman modulation amplitudes have to be applied to improve the sensitivity. Such amplitudes often produce microphonic noise in the cavity and cause an uncertainty in the orientation selection in single crystal-like ENDOR spectra (Sect. 4.1). A modulation technique which avoids these problems in powder ENDOR studies has been proposed by Hyde et al.[32]. In this scheme the Zeeman modulation is replaced by a 180° modulation of the phase of the microwave signal.

2.3 Field Stability

As mentioned above, in an ENDOR experiment the rf field is swept while the static magnetic field is held at a constant position in the EPR spectrum. For slow sweep rates and narrow EPR lines a device would be desirable which is able to stabilize the ratio of the microwave frequency to the static magnetic field. The applicaiton of a commercially available field/frequency lock system is restricted to a region of ± 6 mT about the DPPH resonance field[33]. In metal complexes with strongly anisotropic EPR spectra, however, this locking-field range is usually not sufficient. For such compounds a stabilization of the field/frequency ratio at *any* static field position between 0.08–1.2 T may be achieved with the MIMARS (*mi*crowave frequency to *ma*gnetic field *r*atio *s*tabilization) system developed by Andrist et al.[34]. In this system a digital feedback allows a stabilization of the field/frequency ratio to one part in 10^6.

2.4 ENDOR-Induced EPR (EI-EPR)

In an EI-EPR spectrum (Sect. 4.2) the intensity of a single ENDOR transition is displayed as a function of B_0. Proton-EI-EPR spectra of metal complexes with a width > 10 mT cannot usually be recorded without correcting the offset between the applied radio frequency and the ENDOR resonance condition. Up to now this offset, which is approximately proportional to the change of B_0, was corrected by using digital devices[35,36]. A low cost *analog* set-up which allows an automatic tracking of the observed proton ENDOR line during the B_0 field sweep has been described by Schweiger et al.[37] (Fig. 4). The signal of a self-locking proton magnetometer[38,39] with a frequency output $\nu_p = \mu_N g_p B_0$ is mixed with a frequency or amplitude modulated signal $\Delta\nu$, so that $\nu_p + \Delta\nu$ corresponds to the observed ENDOR resonance. The difference frequency $\nu_p - \Delta\nu$ is suppressed by a high pass filter. While B_0 is swept through the EPR spectrum, the magnetometer frequency follows automatically the Zeeman frequency ν_p of the proton, i.e. the proton ENDOR resonance condition remains approximately fulfilled over the whole B_0 range of the EI-EPR spectrum.

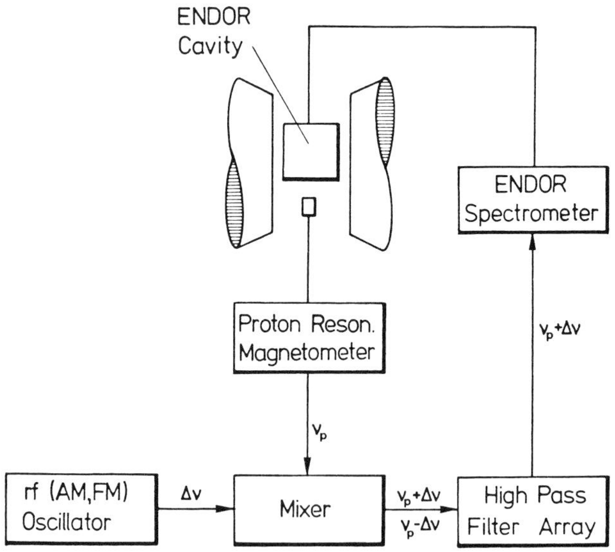

Fig. 4. Block diagram of the EI-EPR spectrometer. (From Ref. 37)

2.5 DOUBLE ENDOR

In DOUBLE ENDOR (Sect. 4.3) the spin system is simultaneously irradiated with two rf fields. The frequencies of both fields have to be set independently from each other. Generation of rf fields by broadband DOUBLE ENDOR instruments are often strong enough for solid state studies at low temperature. In spin decoupling experiments[40] (Sect. 4.4), however, the amplitude of one of the two fields (namely the decoupling field) should be as large as possible, whereas in studies of multiple quantum transitions[41] (Sect. 4.5) two strong rf fields have to be applied.

An X-band high-power DOUBLE ENDOR instrument with two independently tunable resonance circuits has been described by Forrer et al.[42]. The fields RF I and RF II are oriented perpendicular to each other and may be swept between 3–39 MHz and 6–31 MHz, respectively. RF I is generated in a coil which also forms the wall of a TE 011 microwave cavity. Up to 1 kW of rf power may be fed into the impedance matching network consisting of two variable capacitors and the RF I coil. Typically, fields of 2 mT_{rot} (10–20 MHz range) have been achieved in this way[27]. In the RF II coil, which consists of a two-turn loop, an rf power of 150 W generates up to 1 mT_{rot}. The actual design of the DOUBLE ENDOR cavity is shown schematically in Fig. 5.

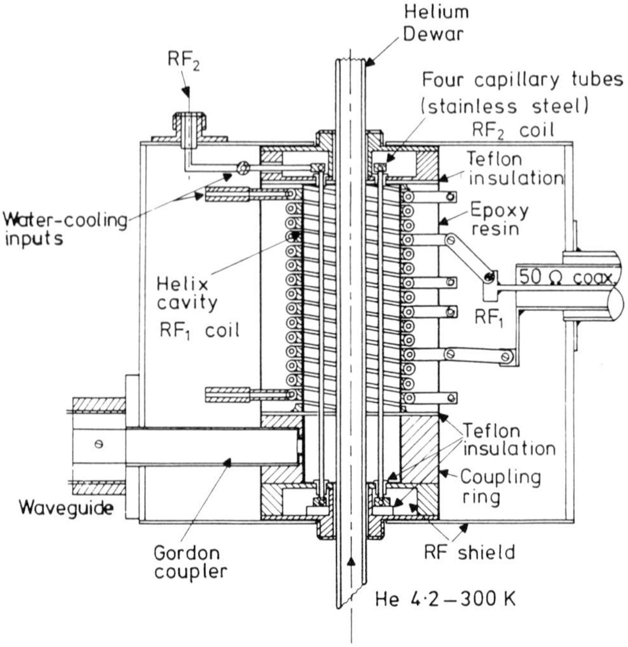

Fig. 5. Schematic diagram of the DOUBLE ENDOR cavity system (X-band). (From Ref. 42)

2.6 ENDOR with Circularly Polarized rf Fields (CP-ENDOR)

The application of circularly polarized rf fields in ENDOR spectroscopy (CP-ENDOR) will be described in Sect. 4.6. Circularly polarized rf fields may be obtained by the vector

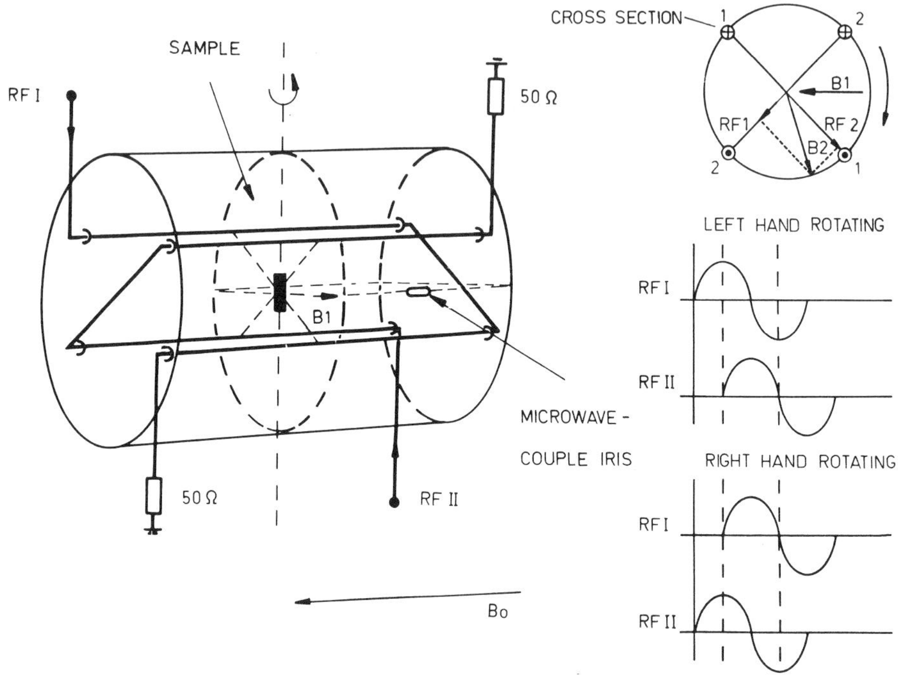

Fig. 6. CP-ENDOR cavity (X-band), operating in the TE 112 mode. $\mathbf{B}_0$: static field, $\mathbf{B}_1$: microwave field, $\mathbf{B}_2$: circularly polarized rf field. (From Ref. 43)

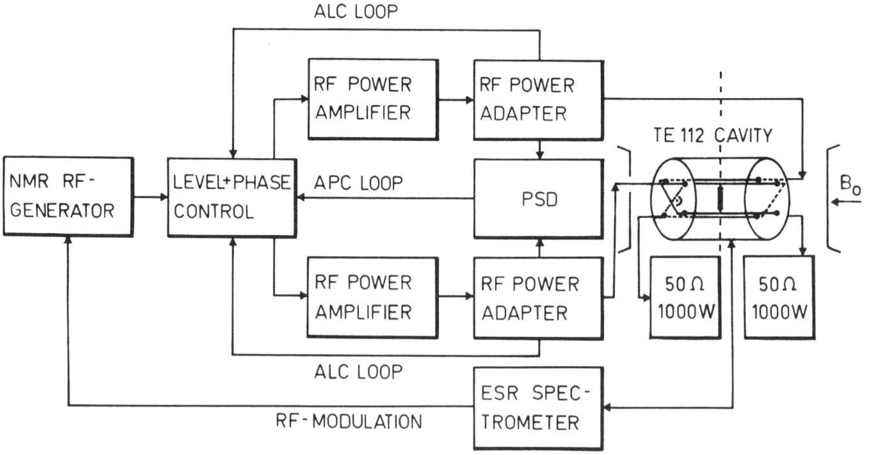

Fig. 7. Block diagram of the CP-ENDOR spectrometer. Frequency range of automatic phase and level control: 2–50 MHz, rf fields 0.08 mT_{rot}. (Adapted from Ref. 43)

sum of two linearly polarized fields $\mathbf{B}_x(t) = \mathbf{i}B_{x0}\cos\omega t$ and $\mathbf{B}_y(t) = \mathbf{j}B_{y0}\cos(\omega t + \varphi)$, with $B_{x0} = B_{y0} = B_2$ and $\varphi = +90°$ for a left hand (l.h.) rotating or $\varphi = -90°$ for a right hand (r.h.) rotating field[43]. Thus

$$\mathbf{B}_{cp}\,(\text{l.h.}) = B_2\,(\mathbf{i}\cos\omega t + \mathbf{j}\cos(\omega t + 90°))$$
$$\mathbf{B}_{cp}\,(\text{r.h.}) = B_2\,(\mathbf{i}\cos\omega t + \mathbf{j}\cos(\omega t - 90°))\ . \tag{2.1}$$

The two linearly polarized fields are generated by rf currents in two pairs of wires oriented parallel to $\mathbf{B}_0$ and connected to form two half-loops which cross each other perpendicularly. A cylindrical cavity operating in the TE 112 mode was considered to be most suitable for this purpose. A schematic diagram of the cavity is shown in Fig. 6.

For geometrical reasons the degree of circular polarization of the rf field is 100% only along the cylinder axis of the cavity. However, it has been shown that for samples with diameters <6 mm the contribution of the counter-rotating component to the nuclear transition probability is less than 1%[43].

The block diagram of the rf part of the CP-ENDOR spectrometer is shown in Fig. 7. To generate rf fields of about 0.08 mT_{rot}, two rf amplifiers with an output power up to 600 W are used. The rf power for each channel is fed through 50 Ω coaxial cables of equal length to the half-loops, and to the 50 Ω loads. A VSWR of about 1/1.2 is achieved with this set-up. The automatic level (ALC) and phase control (APC) units stabilize the two rf amplitudes and hold the phase shift $\varphi = \pm 90°$ constant during the sweep of the radio frequency. Both the phase and the amplitude are measured with high precision current-to-voltage converters (CVC). The output from the CVC's is limited and fed into a broadband phase sensitive detector, allowing a precise phase measurement over the full power and frequency range. A level stability within 1% and a phase tracking error $< \pm 1°$ has been achieved in the range 2 to 50 MHz.

Recently the two techniques DOUBLE ENDOR and CP-ENDOR have been combined (CP-DOUBLE ENDOR) by introducing into the cavity a third half-loop which produces an additional linearly polarized rf field[44].

2.7 Polarization Modulated ENDOR (PM-ENDOR)

In polarization modulated ENDOR spectroscopy (PM-ENDOR)[45], discussed in Sect. 4.7, the linearly polarized rf field $\mathbf{B}_2$ rotates in the laboratory xy-plane at a frequency $f_r << f_m$, where f_m denotes the modulation frequency of the rf carrier. In a PM-ENDOR experiment the same type of cavity, with two rf fields perpendicular to each other, and the same rf level and phase control units used in CP-ENDOR can be utilized. To obtain a rotating, linearly polarized rf field with a constant magnitude B_2 and a constant angular velocity $\Omega = 2\pi f_r$ (f_r typically 30–100 Hz), double sideband modulation with a suppressed carrier is applied to both rf signals. With this kind of modulation the phase of the carrier in each channel is switched by 180° for $\sin\Omega t = 0$. In addition, the phases of the two low-frequency envelopes have to be shifted by 90° with respect to each other. The coding of the two rf signals is shown in Fig. 8.

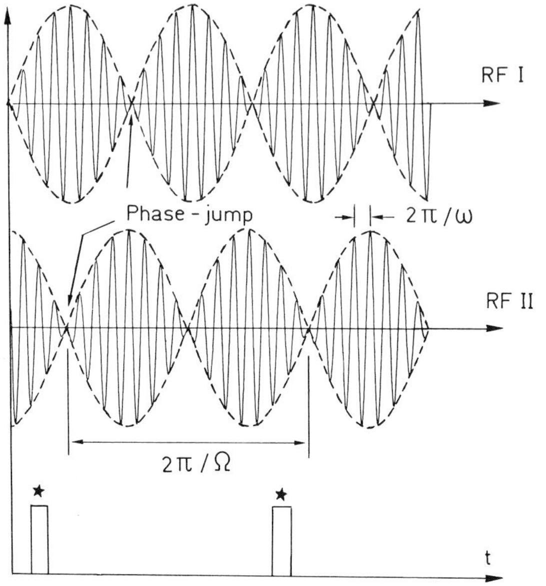

Fig. 8. Coding of the two rf signals, which generate a slowly rotating rf field vector $\mathbf{B}_2$ in the laboratory xy-plane. * mark time intervals for APC. (Ref. 45)

3 Analysis of ENDOR Spectra

In this section analytical expressions for ENDOR transition frequencies and intensities will be given, which allow an adequate description of ENDOR spectra of transition metal complexes. The formalism is based on operator transforms of the spin Hamiltonian under the most general symmetry conditions. The transparent first and second order formulae are expressed as compact quadratic and bilinear forms of simple equations. Second order contributions, and in particular cross-terms between hf interactions of different nuclei, will be discussed for spin systems possessing different symmetries. Finally, methods to determine relative and absolute signs of hf and quadrupole coupling constants will be summarized.

3.1 Spin Hamiltonian

Paramagnetic resonance spectra are usually described by the spin Hamilton operator introduced by Abragam and Pryce[46]. For metal complexes with an effective electron spin $S = 1/2$, the spin Hamiltonian may be expressed as

$$\begin{aligned}\mathcal{H} = {} & \mu_B B_0 \tilde{\mathbf{R}}_{.3} \mathbf{g} \mathbf{S} + \tilde{\mathbf{S}} \mathbf{A}_M \mathbf{I}_M - \mu_N g_M B_0 \tilde{\mathbf{R}}_{.3} \mathbf{I}_M + \tilde{\mathbf{I}}_M \mathbf{Q}_M \mathbf{I}_M + \\ & + \sum_k (\tilde{\mathbf{S}} \mathbf{A}_k \mathbf{F}_k - \mu_N g_k B_0 \tilde{\mathbf{R}}_{.3} \mathbf{F}_k + \sum_\varkappa \tilde{\mathbf{I}}_{k\varkappa} \mathbf{Q}_k \mathbf{I}_{k\varkappa}) \; .\end{aligned} \tag{3.1}$$

In the first row of (3.1) the terms denote the electron Zeeman ($\mathcal{H}_{EZ}$), the hf ($\mathcal{H}_{hfs}$), the nuclear Zeeman ($\mathcal{H}_{NZ}$) and the nuclear quadrupole interaction ($\mathcal{H}_Q$) of the central (metal) ion. The second row represents the hf, the nuclear Zeeman and the nuclear quadrupole interactions for sets of magnetically equivalent ligand nuclei. Each particular set is denoted by the index k, the individual nuclei of set k by $k\varkappa$.

In (3.1) not all tensors are necessarily coaxial or diagonal. If the principal axes system of the $\mathbf{g}$ tensor is chosen as the molecular coordinate system $\{\tilde{\mathbf{e}}^M\}$, $\mathbf{g}$ has diagonal form. The laboratory frame $\{\tilde{\mathbf{e}}^L\}$ is then related to $\{\tilde{\mathbf{e}}^M\}$ by the rotation matrix $\mathbf{R}$ according to

$$\{\tilde{\mathbf{e}}^M\} = \{\tilde{\mathbf{e}}^L\} \, \tilde{\mathbf{R}} \; . \tag{3.2}$$

$\tilde{\mathbf{R}}_{.3}$ in (3.1) is the transpose of the third column of $\mathbf{R}$, and represents a unit vector directed along the static field $\mathbf{B}_0$. More generally $\mathbf{R}_{.i}$, $i = 1, 2, 3$, denotes the ith column of $\mathbf{R}$. This notation proved to be convenient in the description of ENDOR spectra, especially in more complex cases, e.g. in CP ENDOR, PM-ENDOR, DOUBLE ENDOR as well as in spin decoupling experiments (Sect. 4).

3.2 ENDOR Transition Frequencies

For spin Hamiltonians of systems with low symmetry, expressions for ENDOR transition frequencies to second order have been given by several authors[47–55]. In most of these papers the spin Hamilton operator chosen is not general enough to describe the observed ENDOR spectra of transition metal complexes. For instance, some authors consider only one single nucleus[47, 49–52, 54] make use of the assumption $\mathcal{H}_{hfs} \gg \mathcal{H}_{NZ}$[48] or even neglect the term $\mathcal{H}_{NZ}$[51, 52, 54].

For the evaluation of energy levels, ENDOR frequencies and nuclear transition probabilities from the spin Hamiltonian (3.1), we apply the generalized operator transform method, published by Schweiger et al.[55], which is only based on the assumptions $\mathcal{H}_{EZ} \gg \mathcal{H}_{hfs}$ and $\mathcal{H}_{hfs} \gg \mathcal{H}_{Q}$. No restrictions are made on the relative magnitudes of $\mathcal{H}_{hfs}$ and $\mathcal{H}_{NZ}$.

3.2.1 First Order Frequencies

The ENDOR frequencies to first order for a single nucleus with spin I and arbitrary orientation of $\mathbf{B}_0$ are given by[3]

$$\Delta E^{(1)}(m_S,\ m_I + 1 \leftrightarrow m_I) = c(m_S) + \frac{3}{2}\,\overline{Q}_{33}(m_S)(2m_I + 1) \tag{3.3}$$

with

$$c(m_S) = |(\tilde{\mathbf{R}}_{.3}\mathbf{C}(m_S)\tilde{\mathbf{C}}(m_S)\mathbf{R}_{.3})^{1/2}| \tag{3.4}$$

$$\mathbf{C}(m_S) = g^{-1}m_S\mathbf{gA} - \mu_N g_n B_0 \mathbf{I} \tag{3.4'}$$

$$g = |(\tilde{\mathbf{R}}_{.3}\mathbf{g}\tilde{\mathbf{g}}\mathbf{R}_{.3})^{1/2}| \tag{3.5}$$

and

$$\overline{Q}_{33}(m_S) = \tilde{\mathbf{R}}_{.3}\mathbf{C}(m_S)\mathbf{Q}\tilde{\mathbf{C}}(m_S)\mathbf{R}_{.3}/c^2(m_S)\ . \tag{3.6}$$

For each I = 1/2 nucleus, *two* first order ENDOR frequencies, c(1/2) and c(−1/2), are observed. If the **g** tensor and the hfs tensor **A** are *isotropic*, $c(m_S)$ reduces to

$$|m_S a_{iso} - \nu_n|\ , \tag{3.7}$$

i.e., the two ENDOR lines are symmetrically placed about $a_{iso}/2$ or about the nuclear Zeeman frequency $\nu_n = \mu_N g_n B_0$, depending on whether $a_{iso}/2 > \nu_n$ or $a_{iso}/2 < \nu_n$. In the

---

3 It should be noted that in this approach $c(m_S)$ in (3.4) is chosen to be positive. The numbering of the energy levels is therefore the *same* in both m_S-states: increasing m_I-values for increasing energies. Depending on the sign and the magnitude of the tensor elements of **A**, the saturated 'allowed' EPR transition connects energy levels described by the magnetic quantum numbers ($m_S = -1/2,\ m_I \leftrightarrow m_S = 1/2,\ m_I$) *or* ($m_S = -1/2,\ -m_I \leftrightarrow m_S = 1/2,\ m_I$)

case of an *anisotropic* hfs tensor with principal values smaller than ν_n, the two ENDOR frequencies are *not*, in general, located symmetrically to ν_n, but are shifted to higher frequencies, independent of the sign of the principal values of $\mathbf{A}$. The tensor elements of $\mathbf{A}$ may be determined by a non-linear fit of the ENDOR frequencies c(1/2) or c(− 1/2), measured at different static field orientations $\tilde{\mathbf{R}}_{.3}$. The $\mathbf{g}$ tensor in the expressions for $c(m_S)$ has to be evaluated from the orientation dependence of the EPR spectra. If *both* transitions c(1/2) and c(− 1/2) are observed in the ENDOR spectrum, the determination of $\mathbf{A}$ may be simplified using the formula

$$c(-1/2)^2 - c(1/2)^2 = \nu_n g^{-1} \tilde{\mathbf{R}}_{.3} (\mathbf{gA} + \tilde{\mathbf{A}}\tilde{\mathbf{g}}) \mathbf{R}_{.3} \tag{3.8}$$

which is *linear* in $\mathbf{A}$.

For a nucleus with I = 1, the first order ENDOR spectrum consists of four transitions at frequencies

$$\begin{aligned} &c(1/2) \ \pm \frac{3}{2} \overline{Q}_{33}(1/2) \\ &c(-1/2) \pm \frac{3}{2} \overline{Q}_{33}(-1/2) \end{aligned} \tag{3.9}$$

According to (3.3), 4 · I ENDOR transitions will be observed for a nucleus with arbitrary spin I and with an unresolved hf structure in the EPR display. If the hf structure is resolved, however, each m_I-state can be saturated individually and either a four-line ENDOR spectrum (EPR observer: $-I < m_I < I$) or only a two-line ENDOR spectrum (EPR observer: $m_I = \pm I$) will be observed.

For $\mathbf{B}_0$ oriented parallel to one of the principal axes of coaxial $\mathbf{g}$, $\mathbf{A}$ and $\mathbf{Q}$ tensors, the first order transition frequencies in (3.3) reduce to

$$\begin{aligned} &I = \frac{1}{2} : \frac{A_i}{2} \pm \nu_n \\ &I = 1 : \frac{A_i}{2} \pm \nu_n \pm \frac{3}{2} Q_i \\ &I > 1 : \frac{A_i}{2} \pm \nu_n \pm \frac{3}{2} Q_i (2 m_I + 1) , \end{aligned} \tag{3.10}$$

where A_i and Q_i denote the principal values of the tensors $\mathbf{A}$ and $\mathbf{Q}$ along the axis i[4].

A further contribution to the first order ENDOR frequencies arises from the nuclear dipole-dipole interaction $\mathcal{H}_D = \tilde{\mathbf{I}}\mathbf{DK}$ between the two nuclei I and K. The shifts of the ENDOR lines of nucleus I due to $\mathcal{H}_D$ are described by $m_K \overline{D}_{33}(m_S)$, with $\overline{D}_{33}(m_S) = \tilde{\mathbf{R}}_{.3} \mathbf{C}^I(m_S) \mathbf{D} \tilde{\mathbf{C}}^K(m_S) \mathbf{R}_{.3} / c^I(m_S) c^K(m_S)$. In transition metal complexes this interaction is

4 Throughout this paper the principal values of magnetic coupling tensors are denoted by lower indices x, y, z if the corresponding principal axes coincide with the $\mathbf{g}$ tensor axes. In all other cases, indices 1, 2, 3 are used

often observed between the adjacent protons of water molecules, methyl-, or methylene-groups, and may be very helpful for the assignment of hfs tensors to their corresponding nuclei[56–60]. The tensor element $\overline{D}_{33}$ depends on m_S and on the internal fields $\mathbf{B}_e$ caused by the electron at the nuclei I and K. As a consequence, the direction of the maximum dipolar splitting will *not*, in general, be parallel to the internuclear direction. In order to get geometrical information from nuclear dipole-dipole splittings observed in ENDOR spectra, knowledge of the electron **g** tensor and the hfs tensors of the two coupled nuclei is required. The direct nuclear dipole-dipole interaction will be discussed in Sect. 4.4 in more detail.

3.2.2 Second Order Frequencies

First order ENDOR frequencies of nonequivalent nuclei or of pairs of magnetically equivalent nuclei are given by Eq. (3.3) which is derived from the direct product spin base. To obtain correct second order shifts and splittings, however, adequate base functions have to be used. We start the discussion of second order contributions with the most simple case of a single nucleus and will then proceed to more complex nuclear spin systems.

A Single Nucleus

The second order contribution to the transition frequency of a single nucleus (central ion or ligand) with spin I is given by

$$\Delta E^{(2)}(\pm, m_I + 1 \leftrightarrow m_I) = \mp\beta\{a_1(2\,m_I + 1) \mp 2\,a_2 - 2\,a_3(2\,m_I + 1)\} + q', \tag{3.11}$$

with $\beta = (8\mu_B g B_0)^{-1}$,

$$\begin{aligned} a_1 &= \overline{A}_{11}^2 + \overline{A}_{12}^2 + \overline{A}_{21}^2 + \overline{A}_{22}^2 \\ a_2 &= \overline{A}_{12}\overline{A}_{21} - \overline{A}_{11}\overline{A}_{22} \\ a_3 &= \overline{A}_{13}^2 + \overline{A}_{23}^2 \;, \\ \overline{\mathbf{A}} &= \overline{\mathbf{A}}(m_S) = \mathbf{R}^S\mathbf{A}\tilde{\mathbf{R}}^I(m_S) \;, \end{aligned} \tag{3.12}$$

and the second order quadrupole term

$$q' = \{\overline{Q}_{12}^2 + \overline{Q}_{13}^2 + \overline{Q}_{23}^2 + \frac{1}{4}(\overline{Q}_{11} - \overline{Q}_{22})^2\}/2\,c(m_S) \tag{3.13}$$

with $\overline{\mathbf{Q}} = \mathbf{R}^I(m_S)\mathbf{Q}\tilde{\mathbf{R}}^I(m_S)$.

Explicit expressions for the transformation matrices $\mathbf{R}^S$ and $\mathbf{R}^I(m_S)$ determined by

$$\tilde{\mathbf{R}}_{.3}\mathbf{g}\tilde{\mathbf{R}}^S = (001)g \tag{3.14}$$

and $\tilde{\mathbf{R}}_{.3}\mathbf{C}(m_S)\tilde{\mathbf{R}}^I(m_S) = (001)c(m_S)$, respectively, (3.15)

are given in Appendix B, Eqs. (B 2) and (B 3). It is convenient to split (3.11) into different contributions described by the quantities a_i, $i = 1, 2, 3$ (3.12) which produce shifts or splittings directly observable in the ENDOR spectrum. The same type of expressions as given in (3.12) also occur for spin systems with magnetically and geometrically equivalent nuclei.

In many cases the hfs tensor of the central ion is coaxial with the **g** tensor, thus (3.12) reduces to

$$a_1 = \overline{A}_{11}^2 + \overline{A}_{22}^2 , \quad a_2 = -\overline{A}_{11}\overline{A}_{22} , \quad a_3 = \overline{A}_{13}^2 . \tag{3.16}$$

For axial **g** and **A** tensors and $\mathbf{B}_0$ parallel to $g_\parallel$ or $g_\perp$, $\Delta E^{(2)}$ is given by ($A_i/2 > \nu_n > 0$)

$$\Delta E^{(2)}(\pm, m_I + 1 \leftrightarrow m_I) = \mp 4\beta A_\perp^2 (m_I + 1) \tag{3.17}$$

or

$$\Delta E^{(2)}(\pm, m_I + 1 \leftrightarrow m_I) = \mp \beta\{(A_\parallel^2 + A_\perp^2) 2 m_I + (A_\parallel + A_\perp)^2\} , \tag{3.17'}$$

respectively.

In transition metal complexes, proton hfs are normally < 20 MHz so that the corresponding second order contributions, which amount to < 10 kHz, may usually be neglected. For nitrogen ligands, however, the second order corrections produce frequency shifts up to 200 kHz. Since hf interactions of central ions can amount to several hundred megacycles, the terms in $\Delta E^{(2)}$ become very important for a correct description of the ENDOR spectra.

In some compounds the assumption $\mathcal{H}_{EZ} \gg \mathcal{H}_{hfs}$ or $\mathcal{H}_{hfs} \gg \mathcal{H}_Q$ is not fulfilled. In these cases, the second order perturbation approach is no longer accurate enough for the determination of the magnetic parameters. Such a situation has been found, for example, in cobalt and nitrogen ENDOR spectra of *Schiff* base complexes with a large cobalt hfs[61] and with nitrogen hf coupling constants comparable to the ^{14}N nuclear quadrupole interactions[12, 59].

Nonequivalent Nuclei

If two magnetically nonequivalent nuclei I and K are present in the spin system, the transition frequency of nucleus I is shifted by an additional second order term[48, 55]

$$4\beta b_3 m_K = 4\beta(\overline{A}_{13}^I\overline{A}_{13}^K + \overline{A}_{23}^I\overline{A}_{23}^K) m_K , \tag{3.18}$$

where b_3 has an analogous form to the definition of a_3 in (3.12). K denotes any nucleus (central ion or ligand) in the spin state m_K.

The *cross-term* described by (3.18) produces shifts or splittings in the ENDOR spectrum. *Shifts* are observed if the hfs of nucleus K is resolved in the EPR spectrum, i.e. if EPR transitions with different m_K may be used as observers. This is often the case in transition metal complexes with a large hfs of the central nucleus. The shift of nitrogen ENDOR transitions caused by the nitrogen-metal cross-term may amount to more than 1 MHz in copper complexes[62, 63]. *Splittings* or line broadening effects are observed if several hf lines of nucleus K are simultaneously saturated in the EPR spectrum. Such splittings have been observed, for example, in proton ENDOR spectra of γ-irradiated glycine[64].

Since the shifts produced by the cross-terms between ligand and central ion nuclei are often significant, the strikingly broad ENDOR lines ($\lesssim 1$ MHz) observed for transition metal ions[61, 62] may be traced back, at least in part, to unresolved splittings due to second order interactions of numerous ligands with the central ion. ENDOR frequencies up to second order for an $I = 1$ nucleus in the presence of a second ligand nucleus with a spin $K = 1$ are tabulated in Appendix B, Eqs. (B 4).

Two Magnetically Equivalent Nuclei

Many types of transition metal complexes have a center of symmetry, i.e. they contain ligand nuclei which are inversion-related and therefore magnetically equivalent in pairs. For such a complex Schweiger et al.[65] observed a nitrogen ENDOR pattern (Fig. 9) which could not be explained using the second order formulae obtained in the direct product base (B 4). For the description of the observed spectra, a coupled nuclear spin base with $\mathbf{F} = \mathbf{I}_a + \mathbf{I}_b$, $I_a = 1$, $I_b = 1$ and $F = 0, 1, 2$ has been introduced[55, 62]. In this coupling scheme, the quadrupole operator has nonvanishing elements in the off-diagonal block $F = (0, 2)$ of the spin Hamiltonian matrix. Since one of these elements is comparable in magnitude to the difference of the corresponding diagonal elements, the near degeneracy in the second order formula has to be removed by an unitary transformation[62]. The *first order* ENDOR frequencies are again described by the four transitions given in (3.9), i.e. in this approximation the two magnetically equivalent nuclei may be considered as independent from each other.

In second order, however, *eight* ENDOR frequencies are obtained for *each* m_S-state. The transition frequencies tabulated in Appendix B, Eqs. (B 5) are again described by a_1, a_2 and a_3 defined in (3.12). If the hfs is resolved in the EPR spectrum, the number of induced transitions depends on the m_F-value of the saturated line in the EPR quintet. For $m_F = 0$ six transitions, for $|m_F| = 1$ four transitions, and for $|m_F| = 2$ one transition are observed in the ENDOR spectrum of each m_S-state[62].

A typical nitrogen ENDOR spectrum of a copper complex ($Cu(sal)_2$) with two magnetically equivalent ^{14}N nuclei and with the EPR observer at $m_F = 0$ (two sets of six ENDOR lines) is shown in Fig. 9. The pronounced *splitting* of the lines into a doublet structure is described by the term $4\beta a_1$. The splitting of the more intense lines by $4\beta a_3$ is not resolved (see B 5).

Recently similar doublet structures have been observed in other systems with inversion symmetry[58, 66]. Fujimoto et al.[58] used a somewhat different perturbation approach for the explanation of the ^{14}N-ENDOR spectra in copper-doped α-glycine, whereas Brown and Hoffman[66] determined the nitrogen ENDOR frequencies of Cu(TPP) and Ag(TPP) by numerical diagonalization of the spin Hamiltonian matrix for an electron interacting with a single pair of equivalent ^{14}N nuclei.

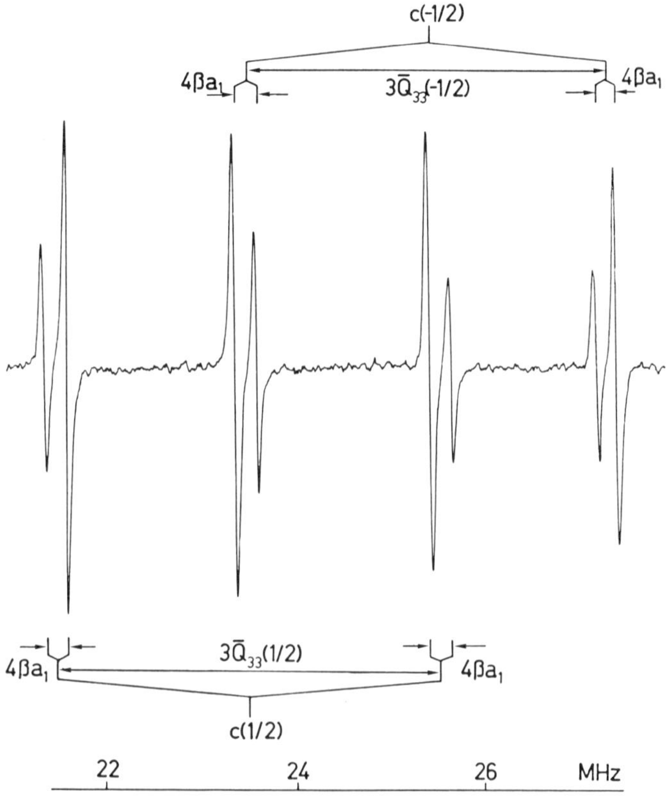

Fig. 9. Second order splittings in spin systems with two magnetically equivalent I = 1 nuclei: Single crystal nitrogen ENDOR spectrum of $Cu(sal)_2$ diluted into $Ni(sal)_2$. (Ref. 62)

Geometrically Equivalent Nuclei

The hfs (or quadrupole) tensors of geometrically (chemically) equivalent nuclei can be transformed into each other by symmetry operations of the point group of the paramagnetic metal complex. For an arbitrary orientation of $\mathbf{B}_0$ these nuclei may be considered as nonequivalent and the ENDOR spectra are described by the simple expressions in (B 4). If $\mathbf{B}_0$ is oriented in such a way that the corresponding symmetry group of the spin Hamiltonian is not the trivial one (C_1 symmetry), symmetry adapted base functions have to be used in the second order treatment for an accurate description of ENDOR spectra. We discuss the C_{2v} and D_{4h} covering symmetry in more detail.

$C_{2v}(e_3^M)$ covering symmetry[67]. For orientations of $\mathbf{B}_0$ in the mirror plane S_{xz}, the symmetry group of the spin Hamiltonian is $\mathcal{G}\{\mathcal{H}\} = C_{2h}(\mathbf{e}_2^M)$. The direct product base of the nuclear spin functions of two geometrically equivalent nuclei reduces to two classes, containing six A-type and three B-type functions, respectively. Second order perturbation theory applied to $\overline{\mathbf{H}} = \mathbf{U}^{\dagger}\mathbf{H}\mathbf{U}$, where $\mathbf{U}$ symmetrizes the base functions of the Hamil-

tonian matrix **H** yields eight ENDOR transitions for each m_S-state (B 6). In these expressions a new cross-term appears, in addition to b_3, namely

$$b_1 = \overline{A}^I_{11}\overline{A}^K_{11} + \overline{A}^I_{12}\overline{A}^K_{12} + \overline{A}^I_{21}\overline{A}^K_{21} + \overline{A}^I_{22}\overline{A}^K_{22} \ , \qquad (3.19)$$

which has an analogous structure to a_1 in (3.12).

D_{4h} covering symmetry. The symmetry of the *M*N_4 unit of Cu(TPP) and Ag(TPP) diluted into (H_2O)Zn(TPP) was found to be essentially D_{4h}[66]. When $\mathbf{B}_0$ is directed along the molecular x or y axis (between the M-N bonds), all four nitrogens are equivalent. For these field orientations, Brown and Hoffman[66] observed a broad ^{14}N-ENDOR pattern with only partially resolved structure. An analytical calculation of the ENDOR frequencies and transition probabilities for these $\mathbf{B}_0$ field orientations is straightforward, but very tedious[68]. To get a better insight into the complexity of the ENDOR spectrum for $\mathbf{B}_0$ oriented along a molecular axis $\mathbf{e}_i^M$, a numerical diagonalization of the full Hamiltonian matrix (four I = 1 nuclei) with $\mathbf{B}_0$ normal to the porphyrin plane has been performed[68]. In Fig. 10 the calculated transition frequencies and intensities for this particular $\mathbf{B}_0$ field orientation are compared with the observed spectrum.

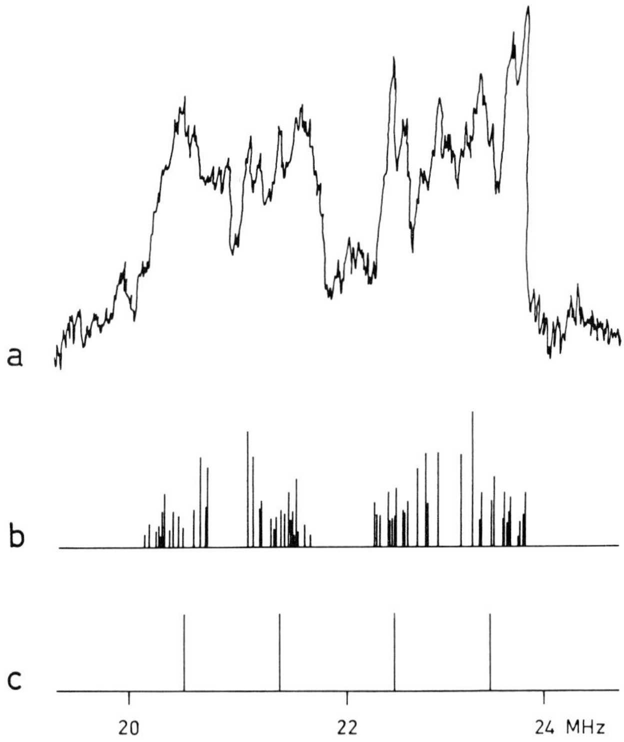

Fig. 10 a–c. Higher order splittings in symmetry planes: Single crystal nitrogen ENDOR spectrum of Cu(TPP) diluted into (H_2O)Zn(TPP) with $\mathbf{B}_0$ normal to the porphyrin plane: B_0 = 327.7 mT. **a)** Observed spectrum. (Adapted from Ref. 66); **b)** Transition frequencies obtained by numerical diagonalization of the full spin Hamiltonian matrix (Four nitrogen nuclei). (Ref. 68); **c)** First order frequencies, (Eq. (3.10))

Noncrossing of Energy Levels

The effects of level noncrossing in ENDOR spectra have been extensively discussed by Schweiger et al.[67]. Two types of noncrossing points have to be distinguished:

1) *proper* noncrossing points, where two levels of the same symmetry would nevertheless be allowed to cross according to analytical expressions of the second order and
2) *improper* noncrossing points where crossing is allowed by symmetry, but in an infinitesimal neighborhood is symmetry-forbidden. Improper noncrossing points occur, for example, at particular $\mathbf{B}_0$ field orientations for which two (or more) nuclei are equivalent. It is this type of noncrossing point which is responsible for the complex ENDOR spectrum depicted in Fig. 10.

Proper noncrossing points may be observed, for example, near $\overline{Q}_{33} = 0$. In systems with two magnetically equivalent nitrogens the typical eight-line pattern (Fig. 9) changes remarkably when the quadrupole coupling changes sign. An experimental example for such a noncrossing region in $Cu(sal)_2$ is given in Fig. 11. The corresponding stick spectrum is obtained by a numerical diagonalization of the full spin Hamiltonian matrix (including the Cu nucleus).

It should be noted that for geometrically equivalent nuclei a complicated ENDOR spectrum may be observed for *arbitrary* orientations of $\mathbf{B}_0$, if the hfs tensors are nearly

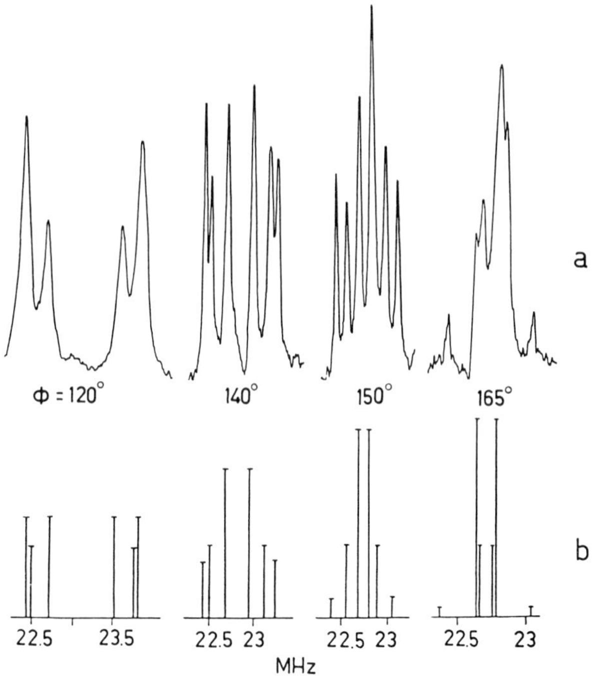

Fig. 11 a, b. Higher order splittings near $\overline{Q}_{33} = 0$ due to noncrossing effects of energy levels: Single crystal nitrogen ENDOR spectrum ($m_S = 1/2$) of $Cu(sal)_2$ diluted into $Ni(sal)_2$ as a function of the orientation of $\mathbf{B}_0$; EPR observer: $m_F = 0$. **a)** Experimental spectrum. At $\varphi = 140°$, the beginning of the repulsion of the energy levels is marked by the splitting of the usual four-line spectrum into a sextet. **b)** Stick spectrum, from numerical diagonalization of the full spin Hamiltonian matrix. (From Ref. 67)

isotropic[67]. In such cases second order expressions in both the direct product and the symmetrized spin base are not suitable for a description of the details of the ENDOR spectrum.

3.3 Transition Probabilities

Nuclear transitions induced by an rf field $B_2(t) = B_2\cos\omega t$ applied along the $\mathbf{x}(\mu = 1)$ or the $\mathbf{y}(\mu = 2)$ direction in the laboratory frame $\{\tilde{\mathbf{e}}^L\}$ are described by the coupling operator

$$\mathcal{H}_1 = B_2(t)\tilde{\mathbf{R}}_{.\mu}(\mu_B\mathbf{g}\mathbf{S} - \mu_N g_M \mathbf{I}_M - \mu_N \sum_k g_k \mathbf{F}_k) \,. \tag{3.20}$$

Application of the generalized operator transform yields for a single nucleus the following nuclear transition probability for *zeroth-order* base functions[45]:

$$W^{(0)} \propto B_2^2 A^2(I, m_I)\{1 - (\tilde{\mathbf{R}}_{.\mu}\tilde{\mathbf{C}}(m_S)\mathbf{R}_{.3})^2/c(m_S)^2\} \tag{3.21}$$

with

$$A^2(I,m_I) = I(I+1) - m_I(m_I - 1) \,.$$

The first term in (3.20) describes the interaction of the oscillating rf field with the electron spin, and contributes only in a first (or higher) order spin base to the nuclear transition probability. For a single nucleus, the *first order* transition probability is given[45] by

$$\begin{aligned} W^{(1)} \propto B_2^2 A^2(I,m_I)\{&\tilde{\mathbf{R}}_{.\mu}\mathbf{C}(m_S)\tilde{\mathbf{C}}(m_S)\mathbf{R}_{.\mu} - (\tilde{\mathbf{R}}_{.\mu}\mathbf{C}(m_S)\mathbf{R}_{.3})^2 \\ &+ (\tilde{\mathbf{R}}_{.\mu}\mathbf{C}(m_S)\mathbf{R}_{.3} - \tilde{\mathbf{R}}_{.\mu}\mathbf{g}\tilde{\mathbf{g}}\mathbf{R}_{.3}\nu_n/g^2)^2 \\ &- (\tilde{\mathbf{R}}_{.\mu}\mathbf{C}(m_S)\tilde{\mathbf{C}}(m_S)\mathbf{R}_{.3} - \tilde{\mathbf{R}}_{.\mu}\mathbf{g}\tilde{\mathbf{g}}\mathbf{R}_{.3}\tilde{\mathbf{R}}_{.3}\mathbf{C}(m_S)\mathbf{R}_{.3}\nu_n/g^2)^2/c(m_S)^2\}/\nu_n^2 \,. \end{aligned} \tag{3.22}$$

Analogous formulae are valid for magnetically equivalent nuclei.

The change of the nuclear transition probability due to the electron spin is usually called *hyperfine enhancement*[2] and has first been discussed by Abragam[69]. The hyperfine enhancement may be considered classically as originating from the modulation of the magnetic field $\mathbf{B}_e$ produced by the electron at the nucleus. The size of $\mathbf{B}_e$ is determined by the hf interaction between the electron and the corresponding nucleus. The field $\mathbf{B}_2^{eff}$ which induces nuclear transitions, is then given by the vector sum of $\mathbf{B}_2$ and the component of $\mathbf{B}_e$ which oscillates with the same frequency as the applied field.

For an isotropic hf interaction a_{iso}, $\mathbf{B}_2^{eff}$ is oriented parallel or antiparallel to $\mathbf{B}_2$ and can be written as $\mathbf{B}_2^{eff} = E\mathbf{B}_2$ with the *enhancement factor*

$$E = \left(1 - \frac{m_S a_{iso}}{\nu_n}\right) . \tag{3.23}$$

For the general case of an anisotropic system the enhancement factor

$$E = \left(\frac{W^{(1)}}{W^{(0)}}\right)^{1/2} \tag{3.24}$$

may easily be calculated from the expressions $W^{(0)}$ and $W^{(1)}$ in (3.21) and (3.22), respectively.

Significant hyperfine enhancements in transition metal complexes are found for central ions with large hf couplings, and for ligand nuclei with small gyromagnetic ratios γ_n. For nitrogen hfs of 40–50 MHz, enhancement factors of $E \simeq 20$ are found at X-band frequencies. In complexes with small ^{14}N hfs (e.g. cobalt and iron compounds), the enhancement factor is near to zero, so that nitrogen ENDOR transitions are often difficult to detect. This is particularly true for one of the two m_S-states. In Co(acacen), for example, the transition probabilities of the low- and high-frequency ^{14}N-ENDOR lines differ by a factor > 10 for most orientations of the field $\mathbf{B}_0$, so that only the high-frequency transitions could be observed[12, 59]. ENDOR intensities of protons which have a large gyromagnetic ratio are not strongly influenced by the hyperfine enhancement. The field $\mathbf{B}_e$, however, plays an important role in proton CP-ENDOR and PM-ENDOR spectra as well as in spin decoupling experiments (Sect. 4).

Nuclear transition probabilities are also influenced by the quadrupole interaction, e.g. for $\mathcal{H}_{hfs} \simeq \mathcal{H}_Q$ transitions with $|\Delta m_I| > 1$ may be induced. Nitrogen ENDOR spectra with $\Delta m_I = \pm 2$ transitions have been observed in several metal complexes with small ^{14}N hf couplings[59, 70, 71].

3.4 Signs of Hyperfine and Quadrupole Coupling Constants

To compare measured magnetic coupling constants with values obtained from theoretical calculations, the signs of the magnetic parameters should be known. In the following, different approaches for the determination of absolute and relative signs of the principal values of hf and quadrupole tensors will be discussed.

Signs of the hyperfine principal values of a single nucleus. According to (3.3) and (3.4) the first order ENDOR frequencies of a single nucleus with spin $I = 1/2$ are given by

$$\begin{aligned} c(m_S) &= |(\tilde{\mathbf{R}}_{.3}\mathbf{C}(m_S)\tilde{\mathbf{C}}(m_S)\mathbf{R}_{.3})^{1/2}| \\ &= \left|\left\{\frac{m_S^2}{g^2}\,\tilde{\mathbf{R}}_{.3}\mathbf{g}\mathbf{A}\tilde{\mathbf{A}}\tilde{\mathbf{g}}\mathbf{R}_{.3} - \frac{m_S\nu_n}{g}\,\tilde{\mathbf{R}}_{.3}(\mathbf{g}\mathbf{A} + \tilde{\mathbf{A}}\tilde{\mathbf{g}})\mathbf{R}_{.3} + \nu_n^2\right\}^{1/2}\right| . \end{aligned} \tag{3.25}$$

The signs of the three hf principal values A_i, $i = 1, 2, 3$, are only determined by the second term in (3.25) which is linear in $\mathbf{A}$. The magnitudes *and* signs of all A_i can therefore be found by a least square fit of the ENDOR data obtained from different orientations of $\mathbf{B}_0$. Since a change of the signs of $\mathbf{A}$ *and* m_S does not alter (3.25), only *relative* signs of A_i can be obtained.

The *absolute* sign of the *isotropic* part of the hf coupling can be determined if the dipolar interaction of a ligand nucleus with the electron in the metal orbitals dominates the other dipolar couplings. In this case the largest principal value of the dipolar part of **A** is positive ($g_k > 0$). This technique has frequently been applied to determine absolute signs of isotropic proton couplings[57, 58, 62].

For nuclei with $A_i \gg \nu_n$, the first term in (3.25) is the dominant one, so that the observed ENDOR frequencies $c(m_S)$ are rather insensitive to the signs of A_i. Since the second order term $2\,a_2\beta$ in (3.11) is dependent on the signs of $\overline{A}_{ij}$ and thus of A_i (3.12), the relative signs of the latter may be found by including higher order contributions in the fitting process.

As mentioned earlier, such a treatment often fails for large hf couplings with second order contributions of several megacycles. Hence, the hf data has to be evaluated by *numerical* diagonalizations of the spin Hamiltonian matrices. This method has been applied to determine the magnitudes and relative signs of the $\mathbf{A}^{Co}$ principal values in Co(acacen)[61]. Since the sign of the largest coupling constant is often known unambiguously from theoretical arguments, the absolute signs of all three principal values may be determined.

Relative signs of the hyperfine splittings of two nuclei. In single crystal ENDOR studies the relative signs between hfs constants of different nuclei may sometimes be determined from the cross-term $4\beta b_3 m_K$ (3.18). If this second order term between nuclei I and K is larger than the ENDOR linewidth, a splitting or shift which is equal for both nuclei within the same m_S-state will be observed in the ENDOR spectrum. This cross-term has been used, for example, in proton ENDOR spectra of AsO_4^{4-} radicals in KH_2AsO_4 to determine the relative signs of the ^{75}As and of the different proton hf coupling constants[72].

Relative signs of hyperfine and quadrupole coupling constants. In an EPR spectrum with resolved hfs the relative signs of hf and quadrupole coupling constants may be determined from the dependence of the ENDOR intensities on the nuclear quantum number m_I of the saturated EPR line[20, 63, 73]. For two magnetically equivalent $I = 1$ nuclei, the signs can be obtained from similar, though more complex, intensity patterns[15, 58, 62].

If the hfs of the nucleus under consideration is not resolved in the EPR spectrum, all nuclear spin states are simultaneously saturated and a sign determination using ENDOR line intensities is not possible. In this case the relative signs may sometimes be determined from second order hf contributions. This method has been applied by DuVarney and Spaeth[74] to determine the sign of the ^{41}K electric quadrupole moment using F centres in KCl.

Sign determinations with the aid of more sophisticated ENDOR techniques will be discussed in the following section.

4 Advanced ENDOR Techniques

In this section we will be concerned with various techniques which supplement the conventional ENDOR experiment. The principles of the different methods as well as their advantages, limitations and applications will be discussed.

Orientation selection in polycrystalline samples and in frozen solutions allows the determination of magnetic coupling parameters along well defined directions of randomly oriented molecules (Sect. 4.1). An improvement in orientation selectivity is obtained by orienting planar paramagnetic compounds in *nematic glasses* (Sect. 4.1). Overlapping EPR or ENDOR spectra of transition metal complexes may be separated by using wide range *ENDOR-induced EPR* (Sect. 4.2) and *DOUBLE ENDOR* (Sect. 4.3), respectively. The technique of *selective decoupling* of nuclear dipole-dipole interactions (Sect. 4.4) and the generation of *multiple quantum coherence* by high rf fields (Sect. 4.5) facilitate the interpretation of ENDOR spectra in many cases. The recent development of *ENDOR with circularly polarized rf fields* (Sect. 4.6) and of *polarization modulated ENDOR* spectroscopy (Sect. 4.7) has been introduced either to reduce the number of the observed ENDOR transitions or to characterize each ENDOR line by additional parameters. Finally, the *envelope modulation* of electron spin echoes and the related *electron spin echo ENDOR* spectroscopy (Sect. 4.8) will be discussed as alternative methods to the conventional ENDOR technique. The different techniques and their applications are summarized in Table 1.

4.1 Orientation Selection in Powders, Frozen Solutions and Nematic Glasses

In ENDOR studies, maximum spectroscopic information is obtained, if the paramagnetic compounds are diluted into diamagnetic single crystal hosts. However, it is often not possible to find suitable host compounds or to grow sufficiently large crystals. This is particularly true with biological systems, e.g. copper or iron proteins. Crystalline powders or frozen solution samples, on the other hand, can normally be prepared more easily. For such randomly oriented compounds, Rist and Hyde[7, 16, 75, 76] have shown that *single crystal-like ENDOR spectra* can be obtained by setting the field B_0 at socalled turning points in the EPR spectrum. Turning points are defined by those magnetic field regions in the powder EPR spectrum which correspond to specific orientations of the molecules with respect to $\mathbf{B}_0$. If only a small portion of the EPR spectrum around a turning point is saturated, an ENDOR signal is observed which corresponds to the single crystal spectrum of this particular orientation.

The applicability of this selection technique depends on the anisotropy and the relative orientation of the various interaction tensors. Frequently occurring situations in transition metal complexes suitable for treatment by this technique are[16, 76]:

Table 1. ENDOR techniques and their applications

Technique	Section	Abbreviation	Experimental	Section	Remarks
Orientation selection	4.1				Determination of hf and quadrupole coupling constants from single crystal-like ENDOR spectra
ENDOR-induced EPR	4.2	EI-EPR	ENDOR resonance frequency correction	2.4	Separation of overlapped EPR spectra and of different orientations in a powder
DOUBLE ENDOR	4.3		two rf fields	2.5	Separation of overlapped ENDOR spectra, determination of relative signs of hf splittings
Nuclear spin decoupling	4.4		two rf fields	2.5	Assignment of ENDOR transitions
Double quantum transitions in ENDOR	4.5		one or two rf fields	2.5	Simplification of the interpretation of ENDOR spectra
ENDOR with circularly polarized rf fields	4.6	CP-ENDOR	circularly polarized rf field	2.6	Reduction of the number of observed ENDOR lines
DOUBLE ENDOR with circularly polarized rf fields	4.6	CP-DOUBLE-ENDOR	one linearly pol. field one circ. pol. field	2.6	Reduction of the number of observed ENDOR lines
Polarization modulated ENDOR	4.7	PM-ENDOR	rf field, lin. pol. rotating in the xy-plane	2.7	Additional labeling of the ENDOR lines, two dimensional displays of the spectrum
Electron spin echo	4.8	ESE	pulsed MW field		Measurement of hf coupling constants
Electron spin echo ENDOR	4.8	ESE-ENDOR	pulsed MW and rf field		Measurement of hf coupling constants

1) The anisotropy of the **g** tensor dominates the anisotropies of all the other magnetic interactions. In this case the turning points at the low- and high-field extremes of the EPR spectrum correspond to molecules oriented such that the magnetic field lies along one of the principal axes of **g**. Many iron complexes, e.g. low- and high-spin ferrimyoglobin, belong to this group.
2) The anisotropy of the **g** and the hfs tensor of the central ion are of the same order of magnitude, but the principal axes of each tensor with the largest anisotropy coincide. Examples are planar cobalt and copper complexes.

Furthermore, the method of orientation selection can only be applied to systems with an electron spin-spin cross relaxation time T_x much larger than the electron spin-lattice relaxation time T_{1e}[77]. In this case, energy exchange between the spin packets of the polycrystalline EPR spectrum by spin-spin interaction cannot take place. If on the other hand $T_x \ll T_{1e}$, the spin packets are coupled by cross relaxation, and a powder-like ENDOR signal will be observed[77]. Since T_{1e}^{-1} is normally the dominant relaxation rate in transition metal complexes, the orientation selection technique could widely be applied in polycrystalline and frozen solution samples of such systems (Sect. 6).

Proper single crystal-like ENDOR spectra can best be obtained by saturating the low- and high-field *flanks* of an EPR spectrum. At these field positions the resolution of the ENDOR spectrum is increased and distortions of the ENDOR lines are minimized[78, 79].

A typical example is illustrated by the five-coordinated cobalt complex Co(salen)py diluted into the corresponding Zn host. For this compound, hf and quadrupole splittings of the pyridine nitrogen have been determined from single crystal-like ENDOR spectra for $\mathbf{B}_0$ along g_x and g_z[80]. Figure 12 shows the EPR powder spectrum and the corresponding ^{14}N-ENDOR spectrum for $\mathbf{B}_0$ along g_x. The well resolved four-line ENDOR pattern in Fig. 12b can be interpreted using the first order formula (3.10) for a nucleus with $I = 1$.

Powder-like ENDOR spectra obtained with arbitrary $\mathbf{B}_0$ orientations show a much less pronounced structure and are usually difficult to interpret[16]. For systems with nearly axial **g** and metal hfs tensors, however, there often exist turning points in the EPR spectrum which correspond to all the $\mathbf{B}_0$ orientations in the *complex plane*. Thus, a setting of the magnetic field at such a turning point results in a powder ENDOR spectrum which is a superposition of the ENDOR spectra arising from all these $\mathbf{B}_0$ orientations. We call it therefore a *two-dimensional* ENDOR spectrum. For a ligand nucleus with $I = 1/2$, the two extreme values of the hfs in the complex plane can immediately be found[78]. For a nucleus with $I \geqslant 1$, however, the evaluation of the two extreme coupling constants of both the hf and the quadrupole interaction, which are not necessarily principal values of **A** and **Q**, requires more sophisticated ENDOR techniques (Sect. 4.3).

In many planar metal complexes it is not possible to record an ENDOR spectrum which only contains contributions from $\mathbf{B}_0$ orientations in the complex plane. This is due to the fact that in the powder EPR spectrum the high- or low-field turning points may arise from *extra absorption peaks*[81], which do not correspond to directions of the principal axes. ENDOR spectra observed near the in-plane region of such a powder EPR spectrum are due to molecules oriented along a large number of $\mathbf{B}_0$ directions (in- and out-of-plane), so that the orientation selection technique is no longer effective.

For cylindrical or disk-like shaped metal complexes the orientation selection may be improved using *nematic glasses* as host compounds. It is well established that the molecules of a liquid crystal in the nematic phase temperature region are oriented in a

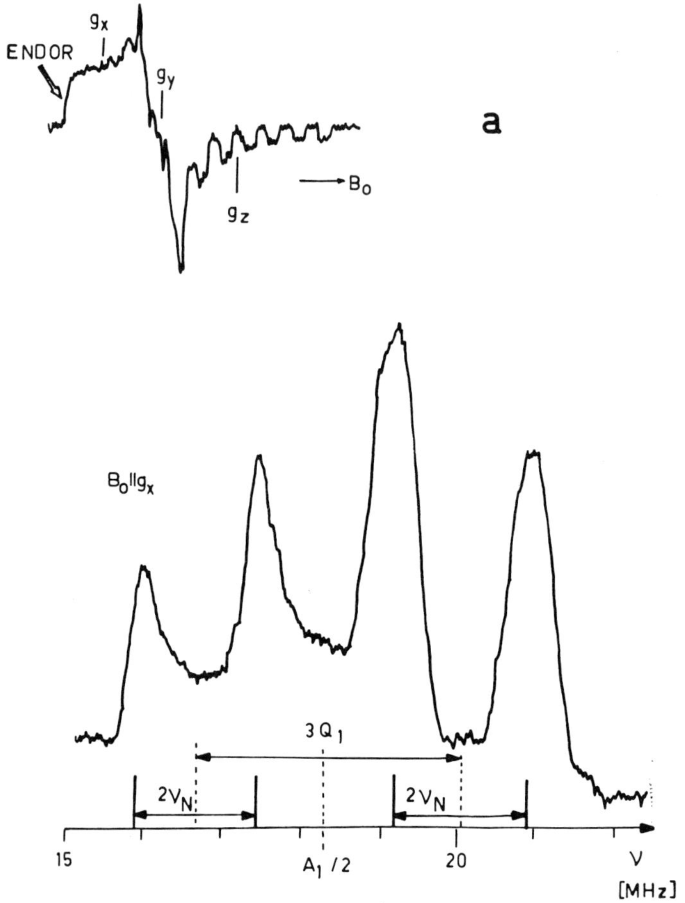

Fig. 12 a, b. Orientation selection in ENDOR. **a)** Powder EPR spectrum of Co(salen)py. Arrow indicates EPR observer; **b)** Single crystal-like ENDOR spectrum of the pyridine nitrogen with $\mathbf{B}_0$ along g_x. (From Ref. 80)

strong external field $\mathbf{B}_0$ along the magnetic field vector[82]. As a consequence, planar paramagnetic solutes are aligned parallel to the solvent molecules. This orientation of the solute molecules can be preserved upon rapidly freezing of the liquid crystal[83]. Thus, in the frozen nematic glass a *two-dimensional powder EPR spectrum* is obtained, consisting of contributions from $\mathbf{B}_0$ only oriented in the complex plane. The out-of-plane directions, and in particular the extra absorption peaks, are often wholly suppressed. This is shown in Fig. 13, which compares the EPR spectra of a planar copper compound in an unoriented (Fig. 13 a) and an oriented (Fig. 13 bc) nematic glass. If the sample is rotated after freezing by 90° (Fig. 13 c), the EPR signal shows features from $\mathbf{B}_0$ oriented perpendicular to the molecular plane which are *enhanced* in intensity compared to the ordinary powder EPR spectrum (improved ENDOR sensitivity along $g_{\|}$!).

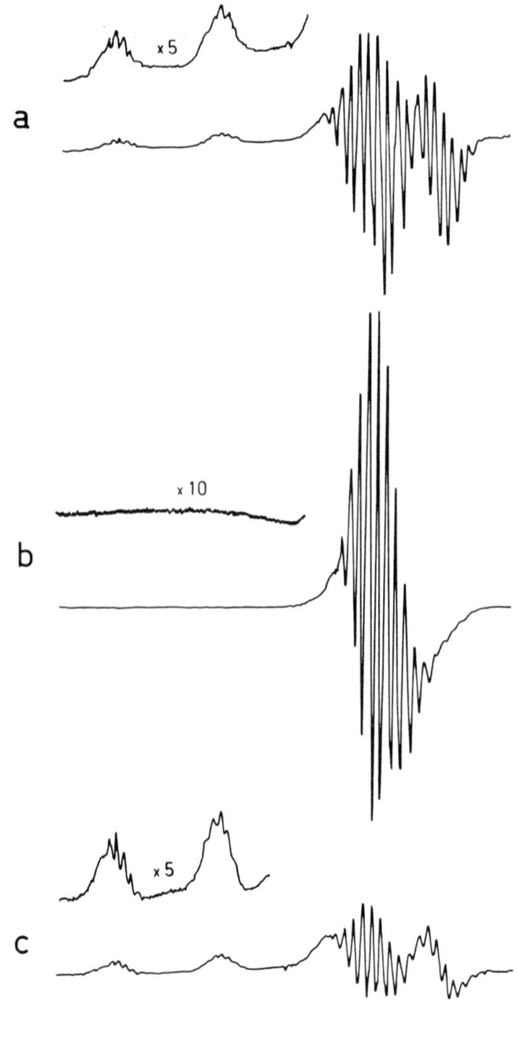

Fig. 13 a–c. EPR spectra of Cu(TPP) diluted into a frozen nematic glass at 10 K (Merck Phase 5). **a)** Unoriented sample. **b)** Frozen nematic glass, oriented along $\mathbf{B}_0$ ($\mathbf{B}_0$ in the complex plane). **c)** Frozen nematic glass, rotated by 90°. (Ref. [84])

Since many planar metal complexes have nearly axially symmetric $\mathbf{g}$ and $\mathbf{A}_M$ tensors, two-dimensional powder ENDOR spectra can easily be obtained from such compounds oriented in nematic glasses[84]. As mentioned, interpretation of this type of spectra will be discussed in Sect. 4.3.

4.2 ENDOR-Induced EPR (EI-EPR)

EI-EPR spectroscopy is a convenient method to separate overlapping EPR spectra originating from different radicals, molecular conformations or magnetic sites. The technique, first demonstrated by Hyde[23], has been applied by several authors, in solutions[85, 86], and in solids[20, 87]. Recently, applications of EI-EPR in single crystal studies have extensively been discussed by Niklas and Spaeth[35].

In an EI-EPR spectrum the variations in the amplitude of an ENDOR line of one particular paramagnetic species is displayed as a function of the static field B_0. Thus, the resulting EI-EPR spectrum corresponds to the EPR signal of *one* species only.

Experimentally, EI-EPR spectra may be obtained in two ways:

1) As the difference spectrum between the conventional EPR signal and the spectrum recorded in the presence of the rf field which induces a nuclear transition in one species. The difference spectrum may be obtained electronically[23, 87] or by computer.
2) A simpler way is to use a frequency modulated rf field and to switch off the Zeeman modulation. With this single coding of the ENDOR line, the EI-EPR spectrum is directly displayed as an absorption signal[20, 35].

Since the nuclear Zeeman frequency ν_n varies as the magnetic field B_0 is swept, a fixed radio frequency *cannot* match the nuclear resonance condition throughout the whole EPR spectrum. For radicals with EPR signals in a field range of about 10 mT and with sufficiently broad ENDOR lines ($\simeq$ 100 kHz), this mismatch results only in intensity variations across the EI-EPR spectrum[87]. However, if the width of the EPR spectrum exceeds 10 mT and if transitions of nuclei with large magnetic moments (^1H, ^{19}F) are used to detect the EI-EPR spectrum, it is necessary to adjust the applied radio frequency according to (3.4). It should be noted that the first order ENDOR frequency $c(m_S, \mathbf{g}, \mathbf{A})$ is *only* linear in B_0 if $\mathbf{g}$ is isotropic and $\mathbf{B}_0$ is oriented along one of the principal axes of the hfs tensor $\mathbf{A}$[37]. In this case, correction of $c(m_S)$ by $\mu_N g_n \Delta B_0$ during the sweep ΔB_0 is accurate. However, this correction, although approximate, is usually sufficient also for many other cases.

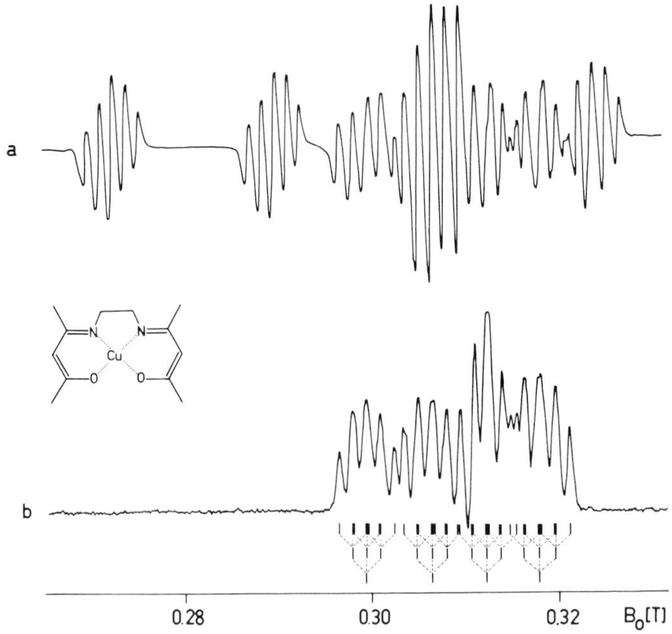

Fig. 14 a, b. EPR and EI-EPR spectra of Cu(acacen) diluted into a Ni(acacen) · $1/2\,H_2O$ single crystal. **a)** EPR spectrum with two overlapping sites (I, II). **b)** EI-EPR spectrum of site II; (hfs of Cu and two ^{14}N nuclei are resolved). (From Ref. 37)

Cu(acacen) diluted into Ni(acacen) · 1/2 H_2O has been chosen as a typical example to demonstrate the separation of magnetically nonequivalent sites in a single crystal by proton EI-EPR. For the specific orientation shown in Fig. 14, the ordinary EPR spectrum of site II is difficult to analyze (Fig. 14a). In the corresponding EI-EPR spectrum (Fig. 14b), a high-frequency proton ENDOR line of this site has been used as an observer. Since site I is completely suppressed in the EI-EPR spectrum, the analysis of the hf data of site II becomes straightforward.

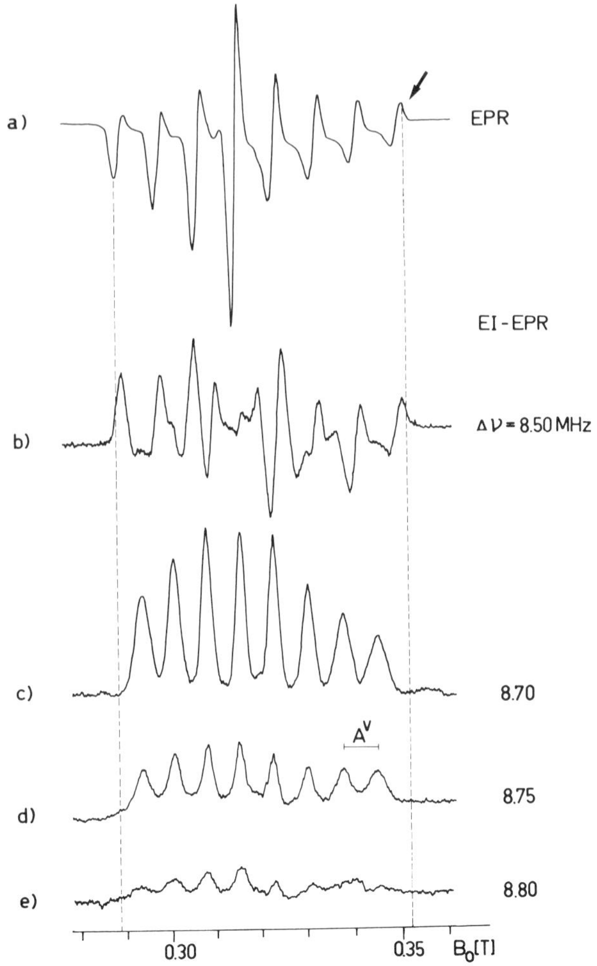

Fig. 15a–e. Single crystal-like EI-EPR spectra from powder samples. **a)** Powder EPR spectrum of $V(bz)_2$ diluted into ferrocene. **b–e)** Sequence of EI-EPR spectra for different $\Delta\nu$ near $A^H_{max}/2$. (From Ref. 37)

EI-EPR has also been applied in powder samples, where *single crystal-like EPR spectra* can be obtained for B_0 observer fields which correspond to the orientations of principal values of ligand hfs tensors[37]. This is illustrated in Fig. 15 for a powder sample of dibenzene vanadium diluted into polycrystalline ferrocene. The hfs tensors of the twelve geometrically equivalent benzene protons are not coaxial with the **g** and the vanadium hfs tensors. Thus, it is not possible to determine all the principal values of the proton hf interaction using the orientation selection technique described in Sect. 4.1. With EI-EPR, the largest hf value $A^H_{max}(A^H_3)$, which lies approximately along the V-H direction, can be measured by recording a set of EI-EPR spectra as a function of $\Delta\nu$ ($\nu = \nu_p + \Delta\nu$). For $\Delta\nu < A^H_{max}/2$ a powder type proton EI-EPR spectrum is observed which originates from a large number of orientations (Fig. 15 b). For $\Delta\nu$ near $A^H_{max}/2$, however, only orientations of the molecules for which $\mathbf{B}_0$ is approximately parallel to the principal axis A^H_{max} contribute to the spectrum, and a well resolved *single crystal-like EI-EPR signal* is recorded as shown in Fig. 15 c. By further increasing $\Delta\nu$ the intensity of the EI-EPR spectrum drops rapidly (Fig. 15 d, e), indicating that $\Delta\nu$ crosses the value $A^H_{max}/2$. If a single crystal linewidth of about 100 kHz is assumed, a principal value of $A^H_{max} = A^H_3 = 17.5$ MHz is estimated from the sequence of EI-EPR spectra shown in Fig. 15 c–e[37]. The angle between the principal axis A^H_3 and the axis normal to the benzene rings has been calculated as 57° from the vanadium hf coupling constant of the single crystal-like EI-EPR spectrum.

Analogous considerations hold for the smallest principal value $A^H_{min} = A^H_2$.

In metal complexes with a large hf interaction of the central ion, the second order shifts due to cross-terms between ligand nuclei and central ion, described by $4\beta b_3 m_M$ (3.18), may not, in general, be neglected. Since these terms are linear in the magnetic quantum number of the central ion, m_M, the ENDOR resonance condition changes *discontinuously* as B_0 is swept from one central ion hf group to the next[37]. The width of the observable EI-EPR spectrum is thus limited and the radio frequency has to be readjusted for each m_M-state. Such second order shifts have to be considered, for example, in nitrogen EI-EPR spectra of copper complexes. Because of the small magnetic moment of ^{14}N, a correction of the mismatch between the pumping frequency and the *first order* ENDOR frequency $c(m_S)$ is usually not necessary for this type of nuclei.

Disregarding incorrect matching of the ENDOR resonance condition, the line shape of the EI-EPR spectrum is only identical to that of the EPR spectrum if[35]: (a) the induced ENDOR transition belongs to an $I = 1/2$ nucleus, (b) only cross relaxation processes of the type $(m_S, m_I) \leftrightarrow (m_S - 1, m_I \pm 1)$ occur, (c) no relaxation takes place between different m_I-states within a given m_S-manifold.

For nuclei with $I \geqslant 1$, the two ENDOR transitions $c(\pm 1/2)$ are split by the quadrupole interaction. As a consequence, only those EPR transitions which have a level in common with the induced ENDOR transition are observed in the EI-EPR spectrum. This selection of the nuclear spin states reduces the number of lines in the EI-EPR spectrum and allows the determination of the relative signs of hf and quadrupole interactions[20, 35].

4.3 DOUBLE ENDOR

In electron-nuclear-nuclear resonance (DOUBLE ENDOR) the change in signal intensity of a single ENDOR transition is observed (RF I) while a second rf field (RF II) is swept through the ENDOR spectrum. Initially, DOUBLE ENDOR (or TRIPLE resonance) was introduced by Cook and Whiffen[88] to determine relative signs of hf coupling constants in single crystals. However, the technique can also be applied to separate overlapping ENDOR signals of magnetically distinct molecules[87].

Investigation of radicals in solutions with DOUBLE ENDOR was proposed by Freed[89] and was first performed by Möbius and coworkers[90–92]. This field was recently covered by a review article[93].

The determination of relative signs of hf coupling constants is illustrated by the energy level diagram in Fig. 16 for a spin system with S = 1/2 and two nonequivalent nuclei with $I_a = I_b = 1/2$. Associated with each EPR line, two low- and two high-frequency ENDOR transitions are observed. The level populations which correspond to saturation of the electron transition (4–8) and the nuclear transition (3–4) are shown in Fig. 16a. In the ENDOR experiment the signal intensity is roughly proportional to the population difference ε between the two levels 3 and 4. In DOUBLE ENDOR this difference is decreased to $2\varepsilon/3$ for transition (2–4) and increased to $4\varepsilon/3$ for transition (7–8), i.e. the intensity of the observed ENDOR transition is *reduced* or *enhanced*, depending on whether the transition induced by the second rf field corresponds to the *same* or to a *different* m_S-state. This change in intensity can therefore be used to determine the relative signs of the two hyperfine coupling constants A_a and A_b.

Both the site separation and the determination of the relative signs of nitrogen ENDOR transitions are demonstrated on the copper complex Cu(acacen) diluted into a Ni(acacen) · 1/2 H_2O single crystal[68]. In the ^{14}N-ENDOR spectrum shown in Fig. 17a, the EPR transitions of both sites (Sect. 4.2) have been saturated simultaneously. If the intensity change of a proton ENDOR line of one site is recorded during the sweep of the second rf field, the DOUBLE ENDOR spectrum shown in Fig. 17b is observed. This spectrum consists of the eight nitrogen ENDOR transitions from only one site (two nitrogen nuclei), and yields immediately the relative signs of the nitrogen hf coupling constants with respect to the sign of the coupling constant of the pumped proton transition.

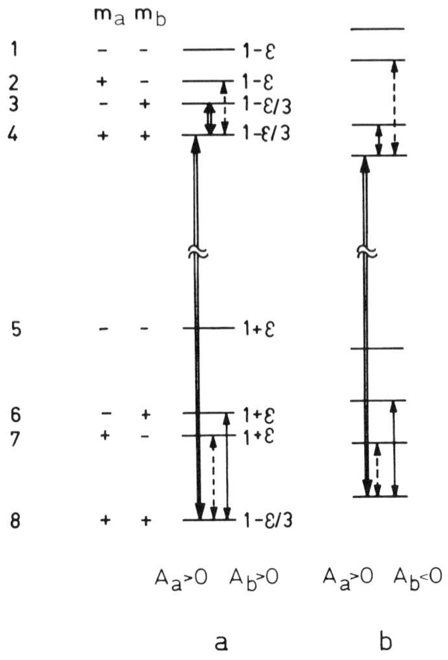

Fig. 16a, b. Energy level diagram for an $S = 1/2, I_a = 1/2, I_b = 1/2$ spin system with $|\nu_n|>|A_a/2|, |A_a|>|A_b|$. **a)** $A_a > 0, A_b > 0$. Level populations when the (4–8) electron transition and the (3–4) nuclear transition $c_a(1/2)$, (*double lines*) are saturated. *Full line*: $c_a(-1/2)$, *dashed lines*: $c_b(\pm 1/2)$. **b)** $A_a > 0$, $A_b < 0$. (Adapted from Ref. 4)

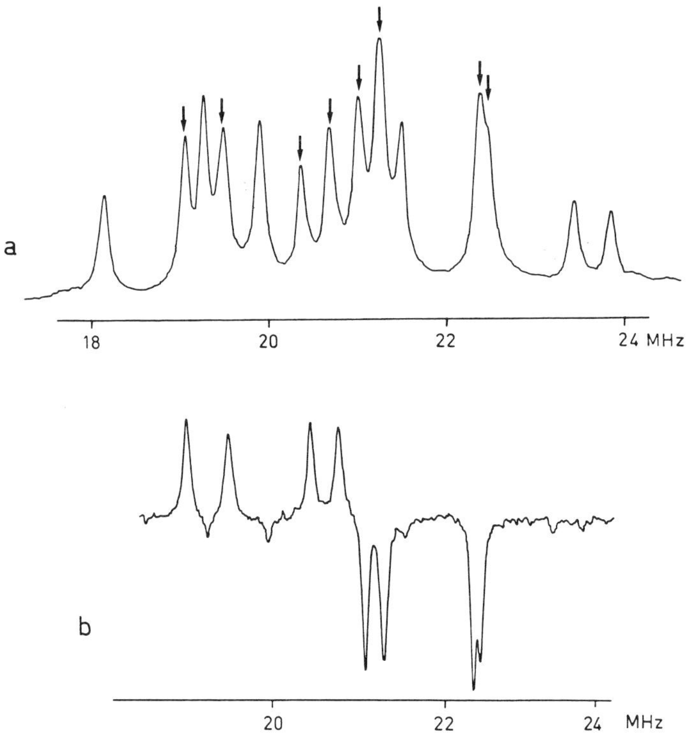

Fig. 17 a, b. Nitrogen DOUBLE ENDOR spectrum of Cu(acacen) diluted into a Ni(acacen) · 1/2 H_2O single crystal. **a)** ENDOR spectrum with two overlapping sites (two sets of eight nitrogen lines). Arrows indicate transitions of the site observed in **b**). **b)** DOUBLE ENDOR spectrum of one site; ENDOR observer: proton ENDOR line at 16 MHz. (From Ref. 68)

As mentioned in Sect. 4.1, the ordering of planar metal complexes in nematic glasses is a convenient method to obtain two-dimensional powder ENDOR spectra. In the following example, the determination of the two principal values of the nitrogen hf and quadrupole coupling constants in the complex plane of Cu(TPP) oriented in a nematic glass will be illustrated. The EPR spectrum of Cu(TPP) with $\mathbf{B}_0$ parallel to the complex plane is shown in Fig. 13b. The corresponding two-dimensional powder nitrogen ENDOR spectrum in Fig. 18 ranges from 19 to 28 MHz and is only poorly resolved. Obviously, it does not allow the determination of the principal values of $\mathbf{A}^N$ and $\mathbf{Q}^N$ in the complex plane. The low- and high-frequency ends of this spectrum are described by $A_1^N/2 - 3/2\ Q_1^N - \nu_N$ and $A_2^N/2 + 3/2\ Q_2^N + \nu_N$, respectively ($A_1^N < A_2^N$). If the low-frequency end at 19 MHz is observed with the field RF I, only orientations with $\mathbf{B}_0$ parallel A_1^N (and A_2^N[84])) will contribute to the DOUBLE ENDOR spectrum shown in Fig. 18b. The three hidden transitions $A_1^N/2 + 3/2\ Q_1^N - \nu_N$ (same m_S-state as the observed transition) and $A_1^N/2 \pm 3/2\ Q_1^N + \nu_N$ (opposite m_S-state) are recorded with reduced and enhanced intensities, respectively. The principal values of A_1^N and Q_1^N are in good agreement with the single crystal data (marked by arrows) reported by Brown and Hoffman[66]). A_2^N and Q_2^N may be found in a similar way by observing the high-frequency end of the nitrogen ENDOR spectrum. For details the reader is referred to[84]).

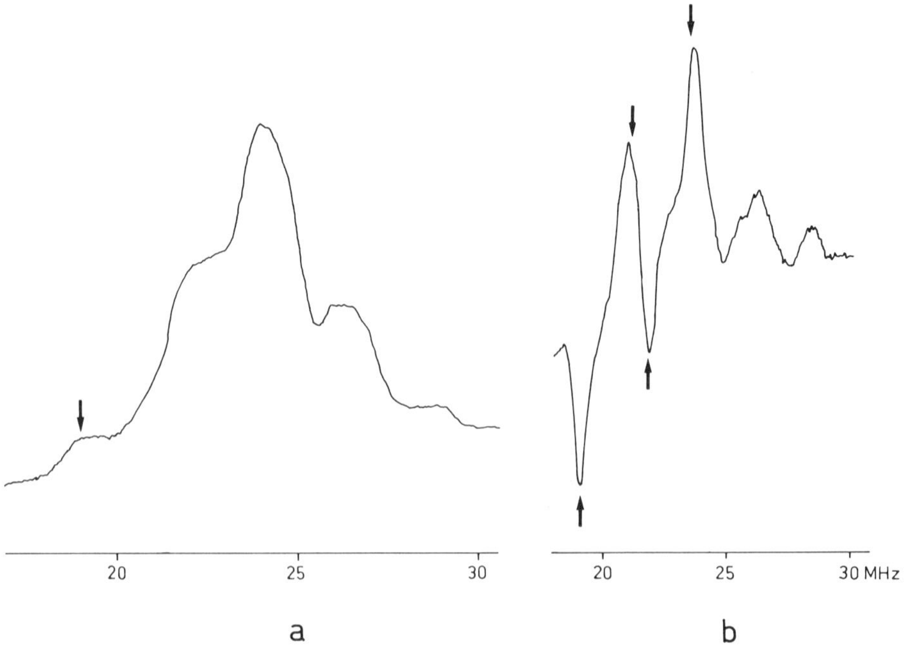

Fig. 18 a, b. DOUBLE ENDOR spectrum of Cu(TPP) in a frozen nematic glass (Merck Phase 5). **a)** Two-dimensional nitrogen ENDOR spectrum with $\mathbf{B}_0$ in the complex plane. ENDOR observer frequency (ν = 19 MHz) used in **b)** is marked by an arrow. **b)** DOUBLE ENDOR spectrum; the corresponding ENDOR frequencies $c_N(\pm 1/2) = A_1^N/2 \pm 3/2\, Q_1^N \pm \nu_N$ obtained from single crystal work[66] are marked by arrows. (Ref. 84)

This example demonstrates that the data A_1^N, Q_1^N, A_2^N, Q_2^N, evaluated from the nematic glass, together with the values A_3^N, Q_3^N, obtained from the single crystal-like ENDOR spectrum, allow the determination of the *full* nitrogen hfs and quadrupole tensors of Cu(TPP) *without* the use of a single crystal.

As mentioned by Baker and Blake[94, 95], DOUBLE ENDOR can be applied for sign determination only if the ENDOR signal is transient or effectively transient[4]. Normally the signals contain transient as well as steady state components. Each modulation cycle freshly excites the ENDOR transition and produces a transient signal which decays to the steady state value. The steady state ENDOR intensity depends on the change of the electron relaxation rate ΔT_{1e} produced by the rf field. A second rf field provides an additional relaxation path between the two levels of the saturated EPR transition and *reduces* the value of ΔT_{1e} for *either* m_S-state. This effect is observed, for example, in single crystal spectra of Co(acacen) at 8 K where all proton and nitrogen transitions of the DOUBLE ENDOR spectrum are detected as *negative* peaks[12], indicating that in this compound T_{1e} is shorter than in the copper complex discussed in Fig. 17 and that the ENDOR signal in Co(acacen) is essentially a steady state one at this temperature. Thus, in such a case, determination of relative signs is no longer possible. However, both the transient and the steady state components cause a change of the ENDOR line intensity, so that *separation* of ENDOR spectra arising from different species can still be achieved using DOUBLE ENDOR, irrespective of the relaxation mechanism.

Another type of DOUBLE ENDOR, called "special TRIPLE", has been introduced by Dinse et al.[90] to study proton hf interactions of free radicals in solution. In a "special TRIPLE" experiment two rf fields with frequencies $\nu_p + \Delta\nu$ and $\nu_p - \Delta\nu$ are swept simultaneously. For systems with $T_{1n} \ll T_{x1}$ this leads to a considerable signal-to-noise improvement and to TRIPLE line intensities which are directly proportional to the number of nuclei with the same hf coupling constant. It should be remembered, however, that in transition metal complexes in the solid state the resonance frequencies are not, in general, symmetrically placed about the free proton frequency ν_p and that the condition $T_{1n} \ll T_{x1}$ is not always fulfilled.

4.4 Nuclear Spin Decoupling in ENDOR Spectroscopy

Nuclear spin decoupling has been widely applied in NMR to simplify complicated spectra in liquid[96] and solid phases[97]. Recently, similar decoupling effects have been observed in single crystal ENDOR spectra with resolved nuclear dipole-dipole couplings between neighboring protons[40]. This interaction can be partially decoupled by a strong second rf field.

We consider the most simple case of a spin system with one electron and two nuclei with spin I = 1/2 and K = 1/2 in a static field $\mathbf{B}_0$ along the laboratory z-axis. A strong pumping rf field with magnitude $2B_2$ and frequency ν_2 is applied along the x-axis and induces transitions of the nucleus K. A weak observing field of magnitude $2B_1$ and frequency ν_1 applied along y causes transitions of the nucleus I.

The spin Hamiltonian for an anisotropic system is then given by

$$\mathcal{H} = \mathcal{H}^\circ + \mathcal{H}' \tag{4.1}$$

with

$$\begin{aligned}\mathcal{H}^\circ = {} & \mu_B B_0 \tilde{\mathbf{R}}_{.3}\mathbf{g}\mathbf{S} + \tilde{\mathbf{S}}\mathbf{A}_I\mathbf{I} - \mu_N g_I B_0 \tilde{\mathbf{R}}_{.3}\mathbf{I} \\ & + \tilde{\mathbf{S}}\mathbf{A}_K\mathbf{K} - \mu_N g_K B_0 \tilde{\mathbf{R}}_{.3}\mathbf{K} + \tilde{\mathbf{I}}\mathbf{D}\mathbf{K}\end{aligned} \tag{4.1 a}$$

and

$$\mathcal{H}' = -2B_2\cos(2\pi\nu_2 t)\tilde{\mathbf{R}}_{.1}(\mu_N g_I \mathbf{I} + \mu_N g_K \mathbf{K} - \mu_B \mathbf{g}\mathbf{S})\ . \tag{4.1 b}$$

The terms in (4.1 a) and (4.1 b) have their usual meaning (Sect. 3).

With the assumptions $B_0 \gg B_2$, $\mu_B g B_0 \gg \nu_2$, $|c_I(m_S) - c_K(m_S)| \gg \max|\overline{D}_{ij}|$ and $|c_I(m_S) - \nu_2| \gg \mu_N g_I B_2$, the *four* ENDOR transitions obtained with the observing frequency ν_1 near $c_I(m_S)$ are given by

$$\begin{aligned} c_{a,b}(m_S) &= c_I(m_S) \pm \frac{1}{2}\{C(+) - C(-)\} \\ c_{c,d}(m_S) &= c_I(m_S) \pm \frac{1}{2}\{C(+) + C(-)\} \end{aligned} \tag{4.2}$$

with

$$C(\pm) = \left\{\left(c_K(m_S) - \nu_2 \mp \frac{1}{2}\bar{D}_{33}\right)^2 + (\mu_N g_K B_2^{eff})^2\right\}^{1/2},$$

where B_2^{eff} denotes the net pumping field introduced in Sect. 3.3. The corresponding ENDOR intensities of these transitions are described by

$$L_{a,b} \propto 1 + \cos\varphi$$
$$L_{c,d} \propto 1 - \cos\varphi \qquad (4.3)$$

with

$$\varphi = \tan^{-1}\frac{\mu_N g_K B_2^{eff}}{(c_K(m_S) - \nu_2 + \bar{D}_{33}/2)} - \tan^{-1}\frac{\mu_N g_K B_2^{eff}}{(c_K(m_S) - \nu_2 - \bar{D}_{33}/2)}.$$

For a derivation of these expressions, the reader is referred to[40]. Equations (4.2) and (4.3) are a generalization of the formulae given by Anderson and Freeman[98] for an isotropic AX nuclear spin system. In contrast to NMR, the applied external decoupling field $\mathbf{B}_2$ in ENDOR is influenced by the internal field $\mathbf{B}_e$ at the nucleus. $\mathbf{B}_e$, which in general does not coincide with the direction of the static field $\mathbf{B}_0$, affects the transition probability induced by the strong rf field $\mathbf{B}_2$ (hyperfine enhancement), and thus influences the decoupling effect. Independent of whether the two involved nuclei are of the same type or not, nuclear dipole-dipole interactions in ENDOR spectra have to be described as a *heteronuclear* coupling. Only for $|c_I(m_S) - \nu_2|$ typically smaller than 100 kHz the interactions are described as a homonuclear coupling and (4.2) and (4.3) have to be revised.

For $\nu_2 = c_K(m_S)$, Eqs. (4.2) reduce to the three transition frequencies

$$c_{a,b}(m_S) = c_I(m_S)$$
$$c_{c,d}(m_S) = c_I(m_S) \pm \frac{1}{2}\{\bar{D}_{33}^2 + (2\mu_N g_K B_2^{eff})^2\}^{1/2}. \qquad (4.4)$$

Figure 19 illustrates the decoupling of the two adjacent amino protons H2 and H3 in $Cu(gly)_2$[40]. The spectral changes of the high-frequency doublet of proton H2 are observed while the transition of proton H3 within the same m_S-state is pumped with the frequency $\nu_2 = c_K(m_S)$ and various $B_2^{eff}(B_2)$ field strengths.

Nuclear spin decoupling in ENDOR may be used to facilitate the analysis of complicated ENDOR spectra. Some of the applications are[40]:

1) Assignment of ENDOR lines to the corresponding pair of nuclei which are coupled by a direct nuclear dipole-dipole interaction.
2) Determination of transition frequencies of hidden lines.
3) Discrimination between splittings which originate from nuclear dipole-dipole couplings and splittings produced by slight deviations from magnetic equivalence.
4) *Determination of relative signs of hyperfine interactions.* In contrast to the sign determination by DOUBLE ENDOR, spin decoupling can also be applied if the ENDOR signals have no significant transient components (Sect. 4.3).

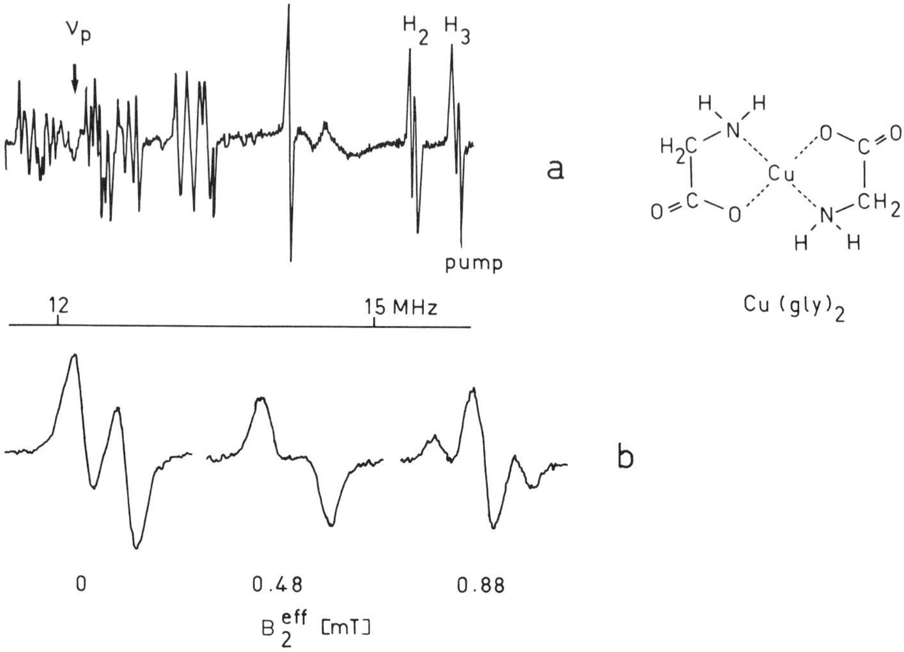

Fig. 19 a, b. Nuclear spin decoupling in ENDOR. **a)** ENDOR spectrum of $Cu(gly)_2$ in α-glycine; ν_p: free proton frequency. **b)** Decoupling sequence of the doublet structure of the proton H2 (I nucleus) for various pumping fields B_2^{eff} at H3 (K nucleus). (Ref. 40)

4.5 Multiple Quantum Transitions in ENDOR

Multiple quantum transitions (MQT) in ENDOR spectra may be observed for nuclei with $I \geqslant 1$ if two or more (rf) photons of the same or of different frequencies combine to produce an ENDOR transition[41, 62, 99–101]. In a MQT the magnetic quantum number m_I changes by $\Delta m_I = \pm n$. The MQT should therefore be clearly distinguished from corresponding 'forbidden' single quantum transitions (SQT) with $|\Delta m_I| > 0$ discussed in Sect. 3.3.

One rf field. The transition probability for a double quantum transition (DQT) between energy levels E_k and E_m generated by one single pumping field of frequency ν is given by[41)]

$$W_{DQT} \propto B_2^4 \frac{|\langle k|I_+|\ell\rangle|^2 |\langle \ell|I_+|m\rangle|^2}{(\nu_{\ell m} - 1/2\,\nu_{km})^2}\, \delta(\nu_{km} - 2\nu) \tag{4.5}$$

with $\nu_{\ell m} = (E_m - E_\ell)/h$ (E_ℓ: intermediate level, Fig. 20 a). W_{DQT} is proportional to the fourth power of the rf field amplitude B_2 and to $(\nu_{\ell m} - 1/2\,\nu_{km})^{-2} = \Delta^{-2}$. To observe a DQT with sufficient intensity, either the amplitude of the pumping field has to be large or the value of Δ small.

Time-dependent perturbation theory shows that the linewidth of an n-quantum transition, generated by a single pumping frequency, should be 1/n of the corresponding

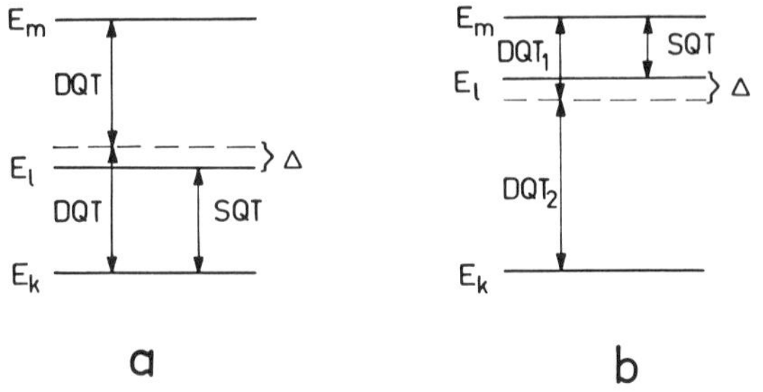

Fig. 20 a, b. Energy level diagram with single (SQT) and double quantum transitions (DQT); **a)** one rf field; **b)** two rf fields

SQT. This is in agreement with the observed linewidth in DQT spectra which are approximately one half (or even less) of the corresponding linewidth of the SQT.

In order to enhance the intensity of a DQT, a possibility varying the parameters B_2 or Δ should be at hand. The field amplitude B_2 is usually limited to $B_2 < 8$ mT for experimental reasons, whereas Δ is determined by the energy level scheme which can only be affected by changing the microwave frequency or the $\mathbf{B}_0$ field orientation. For central ions, the difference in the spacing of three consecutive energy levels is usually determined by both the strong second order hf shifts, and the quadrupole coupling. In this case, it is not always possible to attain sufficiently small Δ-values. In contrast, ligand nuclei have small second order contributions, so that DQT may easily be observed near $\overline{Q}_{33}(m_S) = 0$.

A typical example for the generation of DQT at high B_2 fields in a spin system with two magnetically equivalent I = 1 nuclei (Cu(sal)$_2$) is shown in Fig. 21[41]. The ENDOR spectrum in Fig. 21 a has been recorded with a weak pumping field B_2. Since for this orientation the quadrupole splitting $3\,\overline{Q}_{33}(m_S)$ is only 500 kHz, all six SQT belonging to $m_S = 1/2$ (B 5) can be detected with an EPR observer at $m_F = 0$[62]. For higher pumping fields (Fig. 21 b), four DQT (marked by arrows) are observed. Their positions and intensities are in good agreement with theory[41]. The linewidth of the DQT is only about 1/3 (20 kHz) of the linewidth of the corresponding SQT measured with low B_2 fields. The splitting into doublets of each of the six SQT at high B_2 fields is due to coherence effects[25, 102, 103].

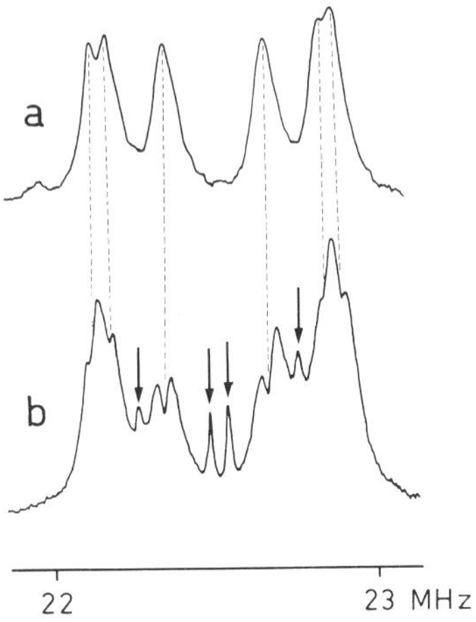

Fig. 21 a, b. DQT in a spin system with two magnetically equivalent I = 1 nuclei (Cu(sal)$_2$, Ref. 62); EPR observer: $m_F = 0$. **a)** Weak pumping field B_2 (0.02 mT$_{rot}$). Only SQT are observed. **b)** Strong pumping field B_2 (0.5 mT$_{rot}$). Four DQT (marked by *arrows*) are observed. SQT are split by coherence effects

DQT generated by a single rf field may be helpful, for example, to elucidate complicated energy level schemes or to assign ENDOR transitions to the corresponding nucleus[41].

Two rf fields[41]. The restriction $E_\ell \simeq (E_m + E_k)/2$ for the intermediate energy level E_ℓ may be dropped if photons of *two different* frequencies generate a DQT. The corresponding transition probability is then given by

$$W_{DQT} \propto (B_2'B_2'')^2 \frac{|\langle k|I_+|\ell\rangle|^2 |\langle \ell|I_+|m\rangle|^2}{(\nu_{k\ell} - \nu_2)^2} \delta(\nu_{km} - \nu_1 - \nu_2) , \tag{4.6}$$

where B_2' and B_2'' denote the amplitudes of the two pumping fields with frequencies ν_1 and ν_2.

To observe a DQT with sufficient intensity, ν_1 and ν_2, with $h(\nu_1 + \nu_2) = E_m - E_k$, may be chosen such that Δ is small for any value of E_ℓ (Fig. 20 b). With two pumping fields and by using appropriate modulation schemes, the SQT can be suppressed and *selective* detection of the DQT is possible[41].

4.6 ENDOR with Circularly Polarized rf Fields (CP-ENDOR)

CP-ENDOR has been introduced by Schweiger and Günthard[104] to reduce the density of ENDOR lines of complicated paramagnetic systems with a large number of interacting nuclei. ENDOR spectra of solutions (liquid or frozen), polycrystalline powders and single crystals can often be simplified remarkably using this technique.

In a spin system, each nuclear spin precesses around its individual effective static field $\mathbf{B}_{eff} = \mathbf{B}_0 + \mathbf{B}_e(m_S)$, (Sect. 3.3). Since the resonance frequency of a nuclear transition is proportional to $\mathbf{B}_{eff}$, ENDOR lines for different types of nuclei may be observed in the same frequency range.

The ENDOR transitions induced by rf fields rotating in the right hand (r.h.) or left hand (l.h.) sense depend on the orientation of $\mathbf{B}_{eff}$ with respect to $\mathbf{B}_0$. For an isotropic spin system the effect of a circularly polarized rf field is readily illustrated in Fig. 22. Two cases (where $a_{iso} > 0$, $g_n > 0$, $g_e = 2$, $S = 1/2$) have to be distinguished:

1) $a_{iso}/2 < \mu_N g_n B_0$ *(Fig. 22 a)*: For both quantum numbers $m_S = \pm 1/2$, $\mathbf{B}_{eff}$ lies parallel to $\mathbf{B}_0$ so that ENDOR transitions are only induced by a l.h. rotating field.
2) $a_{iso}/2 > \mu_N g_n B_0$ *(Fig. 22 b)*: For $m_S = -1/2$, ENDOR transitions are again induced by a l.h. rotating field. For $m_S = 1/2$, however, the orientation of $\mathbf{B}_{eff}$ is reversed and ENDOR transitions will be induced by a r.h. rotating field.

Thus, for a r.h. rotating field, only nuclei which belong to molecules with an electron spin state $m_S = 1/2$ and which have a hf coupling $a_{iso} > 2\mu_N g_n B_0$ will contribute to the ENDOR spectrum.

In a spin system with *anisotropic* $\mathbf{g}$ and $\mathbf{A}$ tensors, the transition probability W_{ba} between two nuclear spin states φ_b and φ_a with energy levels E_b and E_a may be calculated from the coupling operator given in (3.20). For a circularly polarized rf field, $B_2(t)\tilde{\mathbf{R}}_{.\mu}$ in (3.20) has to be replaced by $\tilde{\mathbf{B}}_{cp}(t)\tilde{\mathbf{R}}$ with the l.h. or r.h. rotating field $\mathbf{B}_{cp}(t)$ defined in (2.1). The nuclear transition probability is then given by

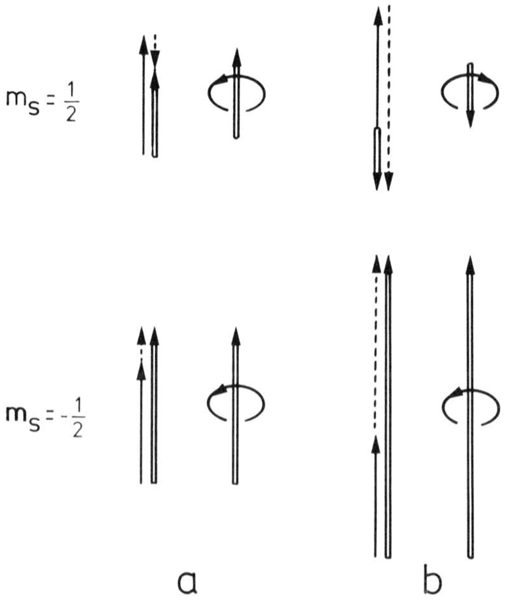

Fig. 22 a, b. Orientation of the effective field $\mathbf{B}_{eff}$ and sense of rotation of the circularly polarized driving field for induction of nuclear transitions. $S = 1/2$, $I = 1/2$, isotropic hf coupling $a_{iso} > 0$, $g_n > 0$. *Solid lines*: external field $\mathbf{B}_0$; *dashed lines:* internal field $\mathbf{B}_e$; *double lines:* effective field $\mathbf{B}_{eff} = \mathbf{B}_0 + \mathbf{B}_e$. **a)** $a_{iso} < 2\mu_N g_n B_0$. **b)** $a_{iso} > 2\mu_N g_n B_0$. (From Ref. 104)

$$\begin{aligned} W_{ba} \propto \delta(\nu_{ba} - \nu)|\langle \varphi_b|B_2\tilde{\mathbf{R}}_{.1}(\mu_B \mathbf{gS} - \mu_N g_n \mathbf{I})|\varphi_a\rangle \\ \mp i \langle \varphi_b|B_2\tilde{\mathbf{R}}_{.2}(\mu_B \mathbf{gS} - \mu_N g_n \mathbf{I})|\varphi_a\rangle|^2 \end{aligned} \tag{4.7}$$

where ν_{ba} and ν denote the resonance frequency and the applied radiation frequency, respectively.

As mentioned in Sect. 3.3, the electron Zeeman term contributes to W_{ba} if φ_b and φ_a are approximated by first order base functions. Explicit first order expressions of W_{ba} for l.h. and r.h. rotating fields, $I = 1/2$ and an isotropic or purely dipolar hfs tensor are given in Table 2.

In the most simple case of an isotropic hfs (Table 2.1) W_{ba} is described by

$$W_{ba} \propto E^2 = \left(1 \mp \frac{a_{iso}}{2\nu_n}\right)^2 .$$

The inversion of $\mathbf{B}_{eff}$ for the low-frequency line takes place at $a_{iso}^{inv} = 2\nu_n = 2\mu_N g_n B_0$. As a consequence the nuclear spin states belonging to $m_S = 1/2$ change their precession direction from l.h. ($a_{iso} < a_{iso}^{inv}$) to r.h. ($a_{iso} > a_{iso}^{inv}$). For $m_S = -1/2$, ENDOR transitions are only observed with a l.h. rotating field.

A different situation is found for a dipolar hfs tensor. The relative nuclear transition probabilities for $\mathbf{B}_0$ parallel to $A_\parallel$ or $A_\perp$ are given in Table 2.2 and 2.3. As in the case of an isotropic hfs tensor, again $\mathbf{B}_{eff}$ is oriented parallel or antiparallel to $\mathbf{B}_0$. Since the enhancement factor E is isotropic for $\mathbf{B}_0$ parallel to $A_\parallel$, the net circularly polarized field rotates in the same sense as the applied field. In the case of $\mathbf{B}_0$ parallel to $A_\perp$, however, the enhancement factor E is *anisotropic*. This anisotropy of E is responsible for the generation of *counter rotating fields* which induce *residual lines*.

Table 2. Relative transition probabilities for circularly polarized fields. $S = 1/2$, $I = 1/2$
1) isotropic hyperfine tensor; $a_{iso} > 0$; 2) dipolar hyperfine tensor; $B_0 \| A_{\|}$; 3) dipolar hyperfine tensor; $B_0 \| A_{\perp}$

	$m_S = \frac{1}{2}$		$m_S = -\frac{1}{2}$	
1)	$a_{iso} < 2\nu_n$	$a_{iso} > 2\nu_n$		
r.h.[a)]	0	$\left[1 - \frac{a_{iso}}{2\nu_n}\right]^2$	0	
l.h.[b)]	$\left[1 - \frac{a_{iso}}{2\nu_n}\right]^2$	0	$\left[1 + \frac{a_{iso}}{2\nu_n}\right]^2$	
2)	$A_{\|} < 2\nu_n$	$A_{\|} > 2\nu_n$		
r.h.	0	$\left[1 + \frac{A_{\|}}{4\nu_n}\right]^2$	0	
l.h.	$\left[1 + \frac{A_{\|}}{4\nu_n}\right]^2$	0	$\left[1 - \frac{A_{\|}}{4\nu_n}\right]^2$	
3)			$A_{\|} < 4\nu_n$	$A_{\|} > 4\nu_n$
r.h.	$\left[\frac{3A_{\|}}{8\nu_n}\right]^2$		$\left[\frac{3A_{\|}}{8\nu_n}\right]^2$	$\left[1 + \frac{A_{\|}}{8\nu_n}\right]^2$
l.h.	$\left[1 - \frac{A_{\|}}{8\nu_n}\right]^2$		$\left[1 + \frac{A_{\|}}{8\nu_n}\right]^2$	$\left[\frac{3A_{\|}}{8\nu_n}\right]^2$

a r.h.: right hand
b l.h.: left hand rotating applied rf field

For an arbitrary orientation of $\mathbf{B}_0$, $\mathbf{B}_{eff}$ will no longer be parallel or antiparallel to $\mathbf{B}_0$. The intensity ratio of transitions induced by l.h. and r.h. rotating fields is then not only determined by the anisotropic enhancement factor but also by the noncoincidence of $\mathbf{B}_{eff}$ and $\mathbf{B}_0$. For proton hfs with $A_{\|}^{H} < 15$ MHz the residual lines induced by a r.h. rotating field will be small, i.e. again an ENDOR spectrum with a reduced number of lines will be observed. In most metal complexes the *dipolar* part of the proton hfs tensors have been found to be below 15 MHz.

The hfs tensors of ligand nuclei in the first coordination sphere of a metal complex are usually dominated by the isotropic interaction, i.e. the transition probabilities may be approximated by the formulae given in Table 2.1 for $a_{iso} > 2\nu_n$.

The following example should illustrate the utility of CP-ENDOR in studies of transition metal complexes. For these compounds the ENDOR frequencies of different types of nuclei often overlap. Moreover, metal complexes frequently contain numerous nuclei of the same type with $A_i > 2\mu_N g_n B_0$ and $I \geqslant 1$, so that suppression of the transitions in one of the two m_S-states simplifies the assignment considerably.

The copper complex $Cu(bipyam)_2(ClO_4)_2$ diluted into the corresponding Zn host crystal[105)] shows an ENDOR spectrum which is due to four magnetically nonequivalent

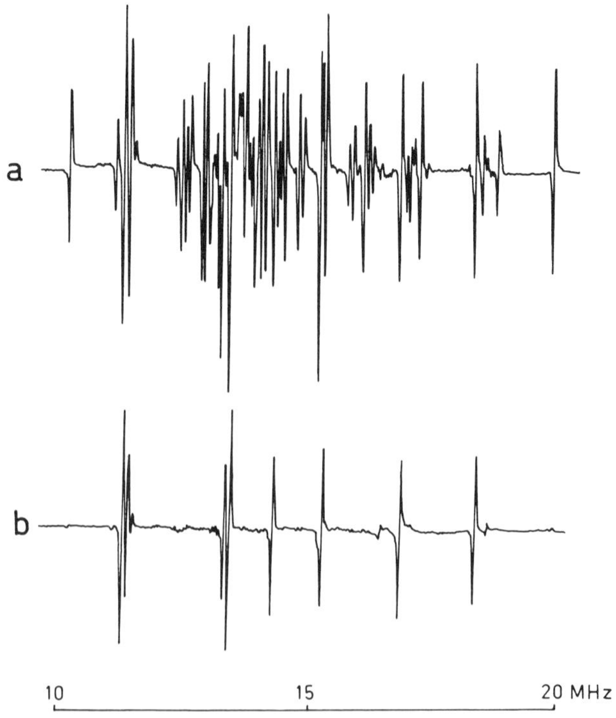

Fig. 23a, b. ENDOR with circularly polarized rf fields (CP-ENDOR). Single crystal ENDOR spectra of $Cu(bipyam)_2$ $(ClO_4)_2$ diluted into $Zn(bipyam)_2(ClO_4)_2$; arbitrary orientation, temperature 20 K. **a)** Conventional ENDOR spectrum (linearly polarized rf field). **b)** CP-ENDOR spectrum: applied rf field right hand rotating. The spectrum is dominated by the eight nitrogen transitions with $m_S A^N > 0$. (From Ref. 104)

^{14}N nuclei (4 sets of 4 transitions) and a large number of proton interactions (Fig. 23 a). It is not possible to discriminate the nitrogen from the proton transitions by their intensities or their linewidths. The corresponding CP-ENDOR spectrum with an rf field rotating in the right hand sense is shown in Fig. 23b. For this polarization only the eight nitrogen transitions with $m_S A^N > 0$ and some weak residual lines are observed. Nevertheless, the spectrum contains all the information about the hf and quadrupole splittings of the four nitrogen nuclei. It is evident from Fig. 23 b that CP-ENDOR drastically simplifies the interpretation of single crystal spectra of this type of compounds. As mentioned, the technique can also be applied to the other states like powders[104] and frozen and liquid solutions.

The combined technique CP-DOUBLE ENDOR[44] derives benefit from both CP-ENDOR and DOUBLE ENDOR methods. It can be applied, for example, to disentangle complicated single crystal ENDOR spectra which are a superposition of several magnetically nonequivalent sites. This is demonstrated in Fig. 24 for the copper complex $[Cu(dien)(bipyam)][NO_3]_2$ diluted into the corresponding Zn compound. Figure 24 a shows a typical ENDOR spectrum which consists of transitions of two overlapping sites. In the CP-ENDOR spectrum, recorded with a r.h. rotating rf field (Fig. 24 b), 16 nitrogen transitions are observed in the frequency range from 6 to 25 MHz (one half of the

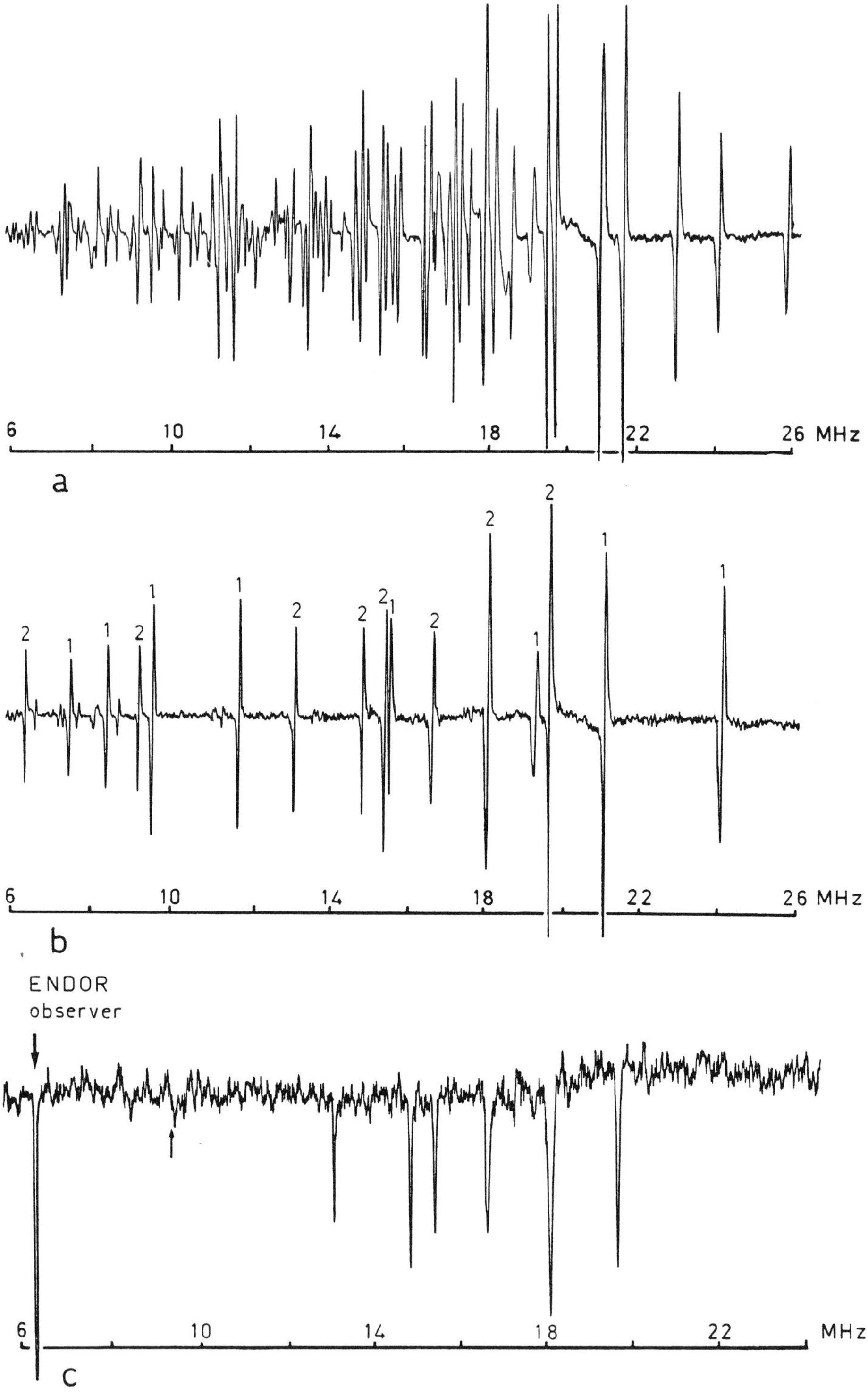

Fig. 24 a–c. DOUBLE ENDOR with a circularly and a linearly polarized rf field. Single crystal spectra of [Cu(dien) (bipyam)][NO_3]$_2$ diluted into [Zn(dien) (bipyam)] [NO_3]$_2$. Arbitrary orientation; temperature 20 K. **a)** Conventional ENDOR spectrum with two overlapping sites. **b)** CP-ENDOR spectrum for a right hand rotating circularly polarized rf field; two overlapping sites. Only nitrogen transitions with $m_S A^N > 0$ are observed. *(1)* and *(2)* denote transitions belonging to site *I* and *II*, respectively. **c)** CP-DOUBLE ENDOR spectrum of site II (swept field: circularly polarized). ENDOR observer (linearly polarized rf field): nitrogen transition at 6.4 MHz of site *II* *(arrow)*. Small arrow marks a weak CP-DOUBLE ENDOR line which could be observed with higher intensity using other ENDOR observers. (Ref. 44)

transitions of two sets of four strongly coupled ^{14}N nuclei). To assign each of these 16 lines to either of the magnetically nonequivalent sites, the intensity change of the transition at 6.4 MHz (say site II) has been observed with a *linearly polarized* rf field, while the *r.h. rotating* rf field was swept through the ENDOR spectrum. The corresponding CP-DOUBLEENDOR spectrum (Fig. 24c) consists only of the eigth nitrogen transitions of site II with $m_S A^N > 0$, so that an unambiguous assignment of all the 16 lines in the CP-ENDOR spectrum is possible.

4.7 Polarization Modulated ENDOR Spectroscopy (PM-ENDOR)

PM-ENDOR, introduced by Schweiger et al.[45], again makes use of the vector property of the effective static field $\mathbf{B}_{eff}$, but in a more selective way than in CP-ENDOR. Figure 25 shows a typical arrangement of the magnetic field vectors $\mathbf{B}_0$, $\mathbf{B}_{eff}$, $\mathbf{B}_2$ and $\mathbf{B}_2^{eff}$ (net pumping field, Sect. 3.3) in the laboratory frame with $\mathbf{B}_0$ along z and the linearly polarized external rf field $\mathbf{B}_2$ at arbitrary direction in the xy-plane. The effective static field vector $\mathbf{B}_{eff} = \mathbf{B}_0 + \mathbf{B}_e(m_S)$ is described by $|\mathbf{B}_{eff}|$ and the polar angles φ and θ.

In ordinary ENDOR, a nuclear transition is characterized by its frequency $\nu = \nu_N g_n B_{eff}$ and its intensity I. Thus, only *one* of three parameters which define the vector $\mathbf{B}_{eff}$, namely $|\mathbf{B}_{eff}|$, is used to distinguish different ENDOR transitions from each other. Since the orientation of $\mathbf{B}_{eff}$ is different for each nuclear transition, the ENDOR lines can be labelled by two new, hitherto not utilized, parameters which are functions of φ and θ. In PM-ENDOR these parameters are determined by using the fact that nuclear transition probabilities are proportional to $\sin^2\alpha$. α denotes the angle between $\mathbf{B}_{eff}$ and $\mathbf{B}_2^{eff}$ (Fig. 25) and is a function of φ, θ and the polarization direction of $\mathbf{B}_2$. Thus, a rotation of $\mathbf{B}_2$ which is restrained to the xy-plane with an angular frequency $\Omega = 2\pi f_r$ (Sect. 2.7) causes a *modulation of the ENDOR line intensity I* which is proportional to the transition probability for weak pumping fields. This intensity modulation, observed at 2Ω[45], is described by $\Delta I = \Delta I_0 \cos(2\Omega t + \varphi_m)$ with $\Delta I_0 = (I_{max} - I_{min})/2$. Each nuclear transition may therefore be characterized by the two new parameters φ_m and $\Delta I_0/\bar{I}$ ($\bar{I} = (I_{max} + I_{min})/2$), which are independent of relaxation mechanisms. For a general discussion of ENDOR with two coherent fields and for explicit expressions of φ_m and $\Delta I_0/\bar{I}$, the reader is referred to[106]. It should be noted that the hyperfine enhancement E essentially

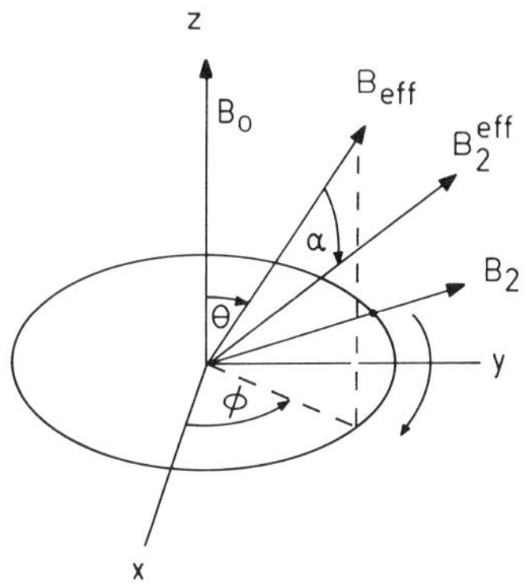

Fig. 25. Magnetic field arrangement of $\mathbf{B}_0$, $\mathbf{B}_{eff}$, $\mathbf{B}_2$ and $\mathbf{B}_2^{eff}$ in the laboratory frame $\{\tilde{\mathbf{e}}^L\}$. $\mathbf{B}_2$ rotates in the xy-plane with an angular frequency Ω. (From Ref. 45)

influences the PM-ENDOR spectrum. For a dipolar **A** tensor, for example, ΔI_0 is maximum (and not zero) for $\mathbf{B}_0$ parallel to $A_\perp$, although $\theta = 0$. In this case the modulation amplitude ΔI_0 and the phase φ_m are fully determined by the anisotropy of E. For strongly coupled nuclei with anisotropic hfs tensors, the angles θ and thus ΔI_0 will, in general, be large. Weak PM-ENDOR signals are expected for nuclei with (nearly) isotropic (θ small, E isotropic) or small hfs (θ and E small).

Since $\Delta I_0/\bar{I}$, φ_m and the ENDOR frequency ν of a particular nucleus are functions of all the elements of the tensor **A**, the full hf information can in principle be determined from the PM-ENDOR data obtained by rotating the single crystal around only *one* arbitrary axis. However, the main benefit from PM-ENDOR compared to conventional ENDOR is its ability to disentangle complicated single crystal ENDOR spectra. The ordinary one-dimensional ENDOR display $I = f(\nu)$ is supplemented by *two-dimensional* displays of the form $\Delta I_{PSD}(\varphi_{PSD}, \nu) = \Delta I_0(\nu)\cos(\varphi_{PSD} - \varphi_m)$ or $\Delta I_0(\varphi_{PSD}, \nu) = \Delta I_0(\nu) \cdot \delta(1 - \cos(\varphi_{PSD} - \varphi_m))$, where φ_{PSD} denotes the phase of the phase sensitive detector. In the latter representations of the data, it is obviously much easier to follow the angular variations of the ENDOR transitions.

The efficiency of the PM-ENDOR technique is illustrated in Fig. 26 by a simple experimental example. Figure 26a shows part of a conventional single crystal proton ENDOR spectrum of Cu(II) diluted into $Mg(NH_4)_2(SO_4)_2 \cdot 6(H_2O)$. Figures 26b, c, which depict the two types of displays mentioned above, show that the parameters φ_m and ΔI_0 are significantly different for each ENDOR transition.

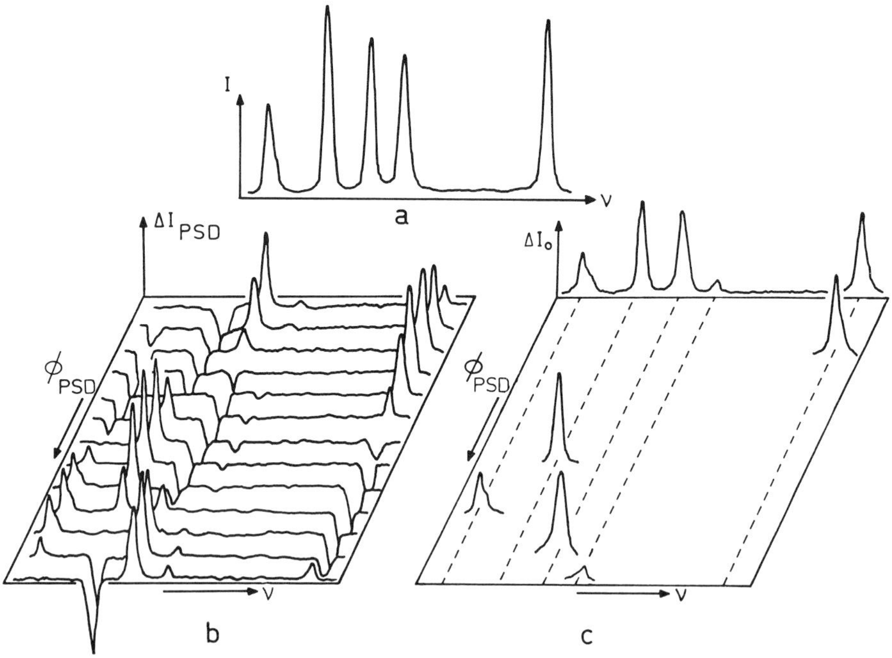

Fig. 26a–c. Polarization modulated ENDOR. **a)** Part of the conventional ENDOR spectrum of Cu(II)-doped $Mg(NH_4)_2(SO_4)_2 \cdot 6H_2O$ (14–16 MHz). **b, c)** PM-ENDOR spectra. Two dimensional displays of ΔI_{PSD} $(\varphi_{PSD}, \nu) = \Delta I_0(\nu)\cos(\varphi_{PSD} - \varphi_m)$ **(b)** and ΔI_0 $(\varphi_{PSD}, \nu) = \Delta I_0(\nu) \cdot \delta(1 - \cos(\varphi_{PSD} - \varphi_m))$ **(c)**. (Ref. 45)

4.8 Electron Spin Echoes and Electron Spin Echo ENDOR

Electron spin echoes (ESE) were first observed in 1958 by Blume[107]. In the last few years this time domain technique has found several interesting applications. For comprehensive summaries on the subject, the reader is referred to the review articles of Mims[108] and Norris et al.[109] and to the monograph of Kevan and Schwartz[110].

ESE envelope modulation. In the context of the present paper the *nuclear modulation effect* in ESE is of particular interest[110, 111]. Rowan et al.[111] have shown that the amplitude of the two- and three-pulse echoes[108] does not always decay smoothly as a function of the pulse time interval τ. Instead, an oscillation in the envelope of the echo associated with the hf frequencies of nuclei near the unpaired electron is observed. In systems with a large number of interacting nuclei the analysis of this modulated envelope by computer simulation has proved to be difficult in the time domain. However, it has been shown by Mims[112] that the *Fourier transform* of the modulation data of a three-pulse echo into the frequency domain yields a spectrum similar to that of an ENDOR spectrum. Merks and de Beer[113] have demonstrated that the display in the frequency domain has many advantages over the parameter estimation procedure in the time domain.

The two techniques, ENDOR and ESE envelope modulation, supplement each other. ESE envelope modulation seems to be more sensitive in detecting nuclear transitions at very low frequencies but is limited in the frequency range by $\gamma_e B_1$, where γ_e denotes the gyromagnetic ratio of the electron and B_1 the microwave pulse amplitude. ENDOR, whose sensitivity increases with frequency, suffers on the other hand from the small transition probability at low frequencies.

As a typical example for the ESE technique, an envelope modulation pattern of Co(acacen) with the corresponding Fourier transform is shown in Figs. 27a, b[114]. The ESE frequencies are in good agreement with those calculated from the high frequency ENDOR data ($A_i^N < 0$, $m_S = 1/2$) for this particular orientation (Fig. 27c). As mentioned in Sect. 3.3, the weak low-frequency nitrogen lines ($m_S = -1/2$) of Co(acacen) could not be observed in ENDOR[59]. In the Fourier transform of the ESE envelope modulation pattern, however, the transitions in the two m_S-states are found to be comparable in intensity. The additional peaks in Fig. 27b are due to sum- and difference-frequencies[114] and possibly to transitions of nitrogen nuclei of the Ni(acacen) host compounds.

The applicability of the ESE envelope modulation technique has been extended by two recent publications[115, 116]. Merks and de Beer[115] introduced a *two-dimensional* Fourier transform technique which is able to circumvent blind spots in the one-dimensional Fourier transformed display of ESE envelope modulation spectra, whereas van Ormondt and Nederveen[116] could enhance the resolution of ESE spectroscopy by applying the *maximum entropy method* for the spectral analysis of the time domain data.

ESE envelope modulation studies of a number of Cu(II) compounds have been reviewed by Peisach[117]. The aim of these investigations was to characterize the chemical environment of the metal binding site in Cu(II) proteins by comparison of the nuclear modulation pattern with those for Cu(II) complexes of known composition.

ESE-ENDOR. In ESE-ENDOR, introduced by Mims[118], a rf pulse in a three-pulse electron spin echo experiment is applied during the time interval between the second and third microwave pulse. The 'ENDOR' spectrum is obtained by monitoring the decrease

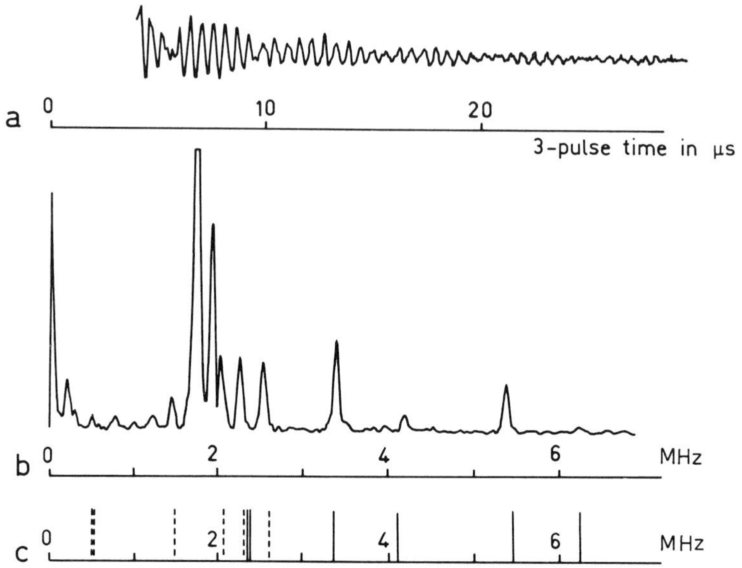

Fig. 27 a–c. Electron spin echo envelope modulation of Co(acacen), temperature 4 K. **a)** Nuclear modulation pattern of Co(acacen) diluted into a Ni(acacen) · 1/2 H_2O single crystal. Crystal setting: rotation axis I, $\varphi = 20°$ [59] (From R. de Beer[114]). **b)** Fourier transform of the nuclear modulation pattern (From R. de Beer[114]) **c)** Stick spectrum: ENDOR frequencies ($\Delta m_N = \pm 1, \pm 2$) calculated from the hfs and quadrupole tensors in Ref. 59 *dashed lines:* $m_S = -1/2$, *full lines:* $m_S = 1/2$

in echo intensity as a function of the radio frequency. For protons, this ENDOR response is modulated with $\cos(2\pi\Delta\nu\tau)$, where τ denotes the time between the first and the second microwave pulse and $\Delta\nu$ the difference between the low- and high-frequency ENDOR lines.

The most important advantage of ESE-ENDOR lies in the fact that it does not critically depend on spin relaxation rates like the conventional cw-ENDOR technique, i.e. ESE-ENDOR might be more sensitive than the latter one. In ESE-ENDOR changes in the echo intensities up to 100% have been reported[109].

It has recently been demonstrated by de Beer et al.[119] that the interpretation of ESE-ENDOR spectra may be simplified by measuring the spectrum at successive fixed values of τ and taking the Fourier transform of the data in the τ-dimension.

5 Interpretation of Hyperfine and Quadrupole Data

The interpretation of magnetic parameters of transition metal complexes has been extensively discussed in the literature. Since a comprehensive summary of the subject is out of the scope of this paper, the reader is referred to a number of textbooks and review articles[2, 120–124]. In this section we shall restrict ourselves to a brief outline of some of the approaches commonly used to interpret hf and quadrupole data obtained from ENDOR spectra.

5.1 Ligand Hyperfine Interactions

The following terms in the Hamilton operator $\mathcal{H}$ give relevant contributions to the hfs tensor of a ligand nucleus N with spin I which interacts with one unpaired electron:

$$\mathcal{H} = \mathcal{H}_{DD} + \mathcal{H}_F + \mathcal{H}_{LS} + \mathcal{H}_{LI} \tag{5.1}$$

with

$$\mathcal{H}_{DD} = P\{[(3\tilde{\mathbf{r}}\mathbf{S})(\tilde{\mathbf{r}}\mathbf{I})/r^5] - \tilde{\mathbf{S}}\mathbf{I}/r^3\} \tag{5.1 a}$$

$$\mathcal{H}_F = \frac{8\pi}{3} P\delta(\mathbf{r}_N)\tilde{\mathbf{S}}\mathbf{I} \tag{5.1 b}$$

$$\mathcal{H}_{LS} = \sum_k \xi(r_k)\tilde{\mathbf{L}}_k\mathbf{S} \tag{5.1 c}$$

$$\mathcal{H}_{LI} = -P(\widetilde{\mathbf{I} \times \mathbf{r}})\mathbf{p}/r^3\,, \tag{5.1 d}$$

where $\mathbf{p}$ is the momentum of the electron, $\mathbf{r}$ the radius vector from the electron to the nucleus N, and $P = \mu_B g_e \mu_N g_n$. The first two terms of (5.1) are the electron-nuclear dipole-dipole operator $\mathcal{H}_{DD}$ described by (5.1 a), and the Fermi contact operator $\mathcal{H}_F$ described by (5.1 b). Both terms contribute to the hf *spin* Hamiltonian $\tilde{\mathbf{S}}\mathbf{A}\mathbf{I}$ in *first order.* The operators $\mathcal{H}_{LS}$ and $\mathcal{H}_{LI}$, defined by (5.1 c) and (5.1 d), describe the spin-orbit coupling (with the one-electron spin-orbit coupling constant $\xi(r_k)$ at the nucleus k) and the interaction of the ligand nucleus with the unquenched part of the orbital magnetic moment, respectively. These two terms contribute in *second order* to the hf spin Hamiltonian[125, 126].

First Order Contributions

Electron-nuclear dipole-dipole coupling. The electron-nuclear dipole-dipole coupling (5.1 a) of the ligand nucleus N is described by

$$A_{ij}^{DD} = P \langle \psi_0 | h_{ij} | \psi_0 \rangle , \tag{5.2}$$

with $$h_{ij} = (3 r_i r_j - \delta_{ij} r^2)/r^5 \tag{5.2 a}$$

and ψ_0 being the MO of the unpaired electron in a molecule with K nuclei. The spin-only coupling tensor $\mathbf{A}^{DD}$ is traceless and symmetric. Expansion of ψ_0 as a linear combination of atomic orbitals $\varphi_{k\varkappa}$ (k = 1, ..., K) yields

$$\begin{aligned} A_{ij}^{DD} &= P \langle \sum_k \sum_\varkappa c_{k\varkappa} \varphi_{k\varkappa} | h_{ij} | \sum_k \sum_\varkappa c_{k\varkappa} \varphi_{k\varkappa} \rangle \\ &= (A_{ij}^{DD})_1 + (A_{ij}^{DD})_{2,1} + (A_{ij}^{DD})_{2,2} + (A_{ij}^{DD})_3 . \end{aligned} \tag{5.3}$$

In (5.3), $(A_{ij}^{DD})_1$ and $(A_{ij}^{DD})_3$ denote the one- and three-center contributions, respectively. In the two-center contribution $(A_{ij}^{DD})_{2,1}$, both atomic orbitals are centered on the same nucleus ($\neq$ N), whereas in $(A_{ij}^{DD})_{2,2}$ one of the atomic orbitals is centered on nucleus N[127]. Contributions $(A_{ij}^{DD})_1$, $(A_{ij}^{DD})_{2,2}$ and $(A_{ij}^{DD})_{2,1}$ are also called "local", "nonlocal" and "distant", respectively[128].

In the ENDOR literature, the tensor $\mathbf{A}^{DD}$ has been treated at several levels of approximation. For ligand nuclei other than protons, the largest coupling generally arises from the one-centre contribution

$$(A_{ij}^{DD})_1 = P \sum_\varkappa \sum_{\bar\varkappa} \langle \varphi_{N\varkappa} | h_{ij} | \varphi_{N\varkappa} \rangle c_{N\varkappa}^* c_{N\bar\varkappa} . \tag{5.4}$$

For protons $(\mathbf{A}^{DD})_1 = \mathbf{0}$, so that the dipolar coupling is determined by the two- and three-center contributions. Recently, Keijzers and Snaathorst[127] have shown that the three-center contributions $(\mathbf{A}^{DD})_3$ should not, in general, be neglected in the computation of anisotropic proton hf coupling constants. In most ENDOR work, however, only the two-center contribution $(\mathbf{A}^{DD})_{2,1}$ ("distant" term) has been considered.

For small distances $R^{(k)}$ ($\lesssim 2$ Å) between the nuclei N and k the two-center integrals in $(\mathbf{A}^{DD})_{2,1}$ can be approximated by expanding (5.1 a) in powers of $r^{(k)}/R^{(k)}$, where $r^{(k)}$ denotes the distance of the electron from the nucleus k[125]. For $R^{(k)} \gtrsim 2.5$ Å, the unpaired electron can be considered to be concentrated at the nuclei k, so that the "distant" contribution may be approximated by the classical electron-nuclear point-dipole formula

$$(\mathbf{A}^{DD})_{2,1} \simeq \mathbf{A}^{DD} = P \sum_{k \neq N} \varrho_k (3\, \mathbf{n}^{(k)} \tilde{\mathbf{n}}^{(k)} - \mathbf{I})/R^{(k)3} , \tag{5.5}$$

where ϱ_k is the spin density on the kth atom. The components of the unit vector $\mathbf{n}^{(k)}$ denote the direction cosines of $R^{(k)}$ in the molecular frame$\{\tilde{\mathbf{e}}^M\}$. The two-center contribution $(\mathbf{A}^{DD})_{2,2}$ is sometimes taken into account to some extent by replacing ϱ_k by Mulliken populations[129].

Since the unpaired electron in transition metal complexes is generally localized near the central ion and the ligand atoms in the first coordination sphere, summation in (5.5) over these nuclei is often sufficient. In this approximated form, the point-dipole model has frequently been applied in ENDOR studies of transition metal complexes to determine the proton positions from their hfs tensors (Sect. 6). In some cases the accuracy of this method has turned out to be significantly higher than that of an X-ray diffraction analysis[62, 130, 131].

Owing to the large spin density ϱ_M at the central ion M, the dipolar coupling (5.5) of the proton is usually dominated by the term

$$\mathbf{A}^{DD} = P\varrho_M(3\,\mathbf{n}^{(M)}\tilde{\mathbf{n}}^{(M)} - \mathbf{I})/R^{(M)3} \, . \tag{5.6}$$

Although (5.6) yields only a rough approximation for the dipolar coupling, it can often be applied to get estimates of $R^{(M)}$ and $\mathbf{n}^{(M)}$ and thus to assign the hfs tensors to the corresponding protons.

The dipolar *splitting* is described by the simple first order expression

$$P\varrho_M(3\cos^2\theta - 1)/R^{(M)3} \, , \tag{5.7}$$

where θ is the angle between $\mathbf{B}_0$ and the line connecting the central nucleus with the proton.

Fermi contact interaction. The coupling constant of the Fermi contact term for the nucleus N has the form[127]

$$\begin{aligned} a^F &= \frac{8\pi}{3} P\langle\psi_0|\delta(\mathbf{r}_N)|\psi_0\rangle \\ &= a_1^F + a_{2,1}^F + a_{2,2}^F + a_3^F \, . \end{aligned} \tag{5.8}$$

If only the one-center contribution a_1^F is considered, (5.8) reduces to

$$a^F \simeq a_1^F = a_{iso} = \frac{8\pi}{3} P\varrho(\mathbf{r}_N = 0) \, , \tag{5.9}$$

where $\varrho(\mathbf{r}_N = 0) = f_s^N\,|\varphi_{N,s}(\mathbf{r}_N = 0)|^2$ denotes the unpaired electron density at the nucleus N (in electrons/angstrom^3).

Second Order Contributions

In transition metal complexes there may be a substantial orbital magnetic moment (reflected by deviations of the g-values from g_e) which leads to an *isotropic* as well as an *anisotropic* contribution to the hf interaction[125, 132]. These contributions arise from a second order term of the form $\langle\psi_0|\mathcal{H}_{LI}|\psi_m\rangle\langle\psi_m|\mathcal{H}_{LS}|\psi_0\rangle$, where ψ_m denotes the MO of the unpaired electron in the excited state m. We consider only the interaction of the ligand nucleus N with the unquenched orbital momentum on the central ion. Expansion of r^{-1} in (5.1d) in powers of $(r^{(M)n}/R^{(M)n+1})$[125] yields, for the leading term

$$\mathcal{H}_{LI} = \mu_B g_n \mu_N (3\,I_\sigma L_\sigma - \tilde{\mathbf{I}}\mathbf{L})/R^{(M)3}\,, \tag{5.10}$$

where σ refers to the metal-ligand direction. The orbital contribution to the spin Hamiltonian of nucleus N is described by $\tilde{\mathbf{S}}\mathbf{A}^L\mathbf{I}$ where the coupling matrix $\mathbf{A}^L$ is given by

$$\mathbf{A}^L = (\mathbf{A}^L)_{2,1} = 2\,\xi\mathbf{\Lambda}\mathbf{T}^d\,. \tag{5.11}$$

In (5.11), $\mathbf{\Lambda}$ and $\mathbf{T}^d$ are defined by

$$\mathbf{\Lambda} = \sum_m \frac{\langle\psi_0|L_i|\psi_m\rangle\langle\psi_m|L_j|\psi_0\rangle}{E_0 - E_m} \tag{5.11 a}$$

and

$$\mathbf{T}^d = \frac{\mu_B\mu_N g_n}{R^{(M)3}} \begin{pmatrix} -1 & & \\ & -1 & \\ & & 2 \end{pmatrix} \tag{5.11 b}$$

and ξ is the effective one-electron spin-orbit coupling parameter. Since only the orbital momentum of the central ion is considered, the MO's ψ_0, ψ_m in (5.11 a) consist solely of atomic orbitals of this ion. Replacement of $2\,\xi\mathbf{\Lambda}$ by $\Delta\mathbf{g} = \mathbf{g} - g_e\mathbf{I}$[133] yields

$$\mathbf{A}^L = \Delta\mathbf{g}\mathbf{T}^d \tag{5.12}$$

which in the molecular frame ($\mathbf{g}$ diagonal) writes

$$\mathbf{A}^L = \Delta\mathbf{g}\mathbf{T} = \Delta\mathbf{g}\,\frac{\mu_B\mu_N g_n}{R^{(M)3}}\,(3\,\mathbf{n}^{(M)}\tilde{\mathbf{n}}^{(M)} - \mathbf{I})\,. \tag{5.12 a}$$

Since (5.12 a) is the product of a diagonal matrix $\Delta\mathbf{g}$ with a full, traceless and symmetric matrix $(3\,\mathbf{n}^{(M)}\tilde{\mathbf{n}}^{(M)} - \mathbf{I})$, $\mathbf{A}^L$ will, in general, be *asymmetric*.

The phenomenon of asymmetric hfs tensors was first discussed by McConnell[134]. Later, Kneubühl[135, 136] proved the existence of asymmetric $\mathbf{g}$ and $\mathbf{A}$ tensors in paramagnetic systems with low symmetry. Evaluation of the asymmetry of $\mathbf{A}$ using EPR and ENDOR spectroscopy has been treated by several authors[132, 137–141]. Recently, low-symmetry effects in EPR have been covered in a comprehensive review article by Pilbrow and Lowrey[142].

Summing the first and second order contributions (5.9), (5.5) and (5.12 a), the *total* hf matrix for a *proton* (ligand atom N) in the point-dipole approximation is given by

$$\mathbf{A} = a^F\mathbf{I} + \mathbf{A}^{DD} + \mathbf{A}^L = \mu_B\mu_N g_p \left\{\frac{8\pi}{3}\,g_e\varrho(\mathbf{r}_N = 0)\mathbf{I} + \frac{g_e\varrho_M}{R^{(M)3}}\,(3\,\mathbf{n}^{(M)}\tilde{\mathbf{n}}^{(M)} - \mathbf{I}) \right.$$
$$\left. + \; g_e \sum_{\substack{k\neq M\\ k\neq N}} \varrho_k \frac{(3\,\mathbf{n}^{(k)}\tilde{\mathbf{n}}^{(k)} - \mathbf{I})}{R^{(k)3}} + \frac{\Delta\mathbf{g}}{R^{(M)3}}\,(3\,\mathbf{n}^{(M)}\tilde{\mathbf{n}}^{(M)} - \mathbf{I})\right\}. \tag{5.13}$$

Due to the inclusion of $\mathbf{A}^L$, $\mathbf{A}$ then becomes *asymmetric*. Since the expression (3.25) for the first order ENDOR transition frequencies contains the term $\tilde{\mathbf{R}}_{.3}(\mathbf{gA} + \tilde{\mathbf{A}}\tilde{\mathbf{g}})\mathbf{R}_{.3}$, which is linear in $\mathbf{A}$, all *nine* matrix elements of the hfs tensor $\mathbf{A}$ can, in principle, be determined by ENDOR. If, in the third term of (5.13), the spin densities ϱ_k ($k \neq N$) can be neglected (ionic model, $\varrho_k = 0$, $\varrho_M = 1$), the hfs tensor $\mathbf{A}$ is described by

$$\mathbf{A} = a^F\mathbf{I} + \mathbf{gT} \ . \tag{5.14}$$

Equation (5.14) amounts to a description of the anisotropic coupling $\mathbf{A}^{DD} + \mathbf{A}^L$ originating from the interaction of an electron spin magnetic moment $\boldsymbol{\mu} = \mu_B \mathbf{gS}$ with the proton magnetic moment[134]. The isotropic part of $\mathbf{gT}$ which is usually called the *pseudo contact interaction*[143], is only zero for isotropic $\mathbf{g}$ tensors. Since in (5.14) the $\mathbf{g}$ tensor is known from EPR data, the remaining parameters which describe $\mathbf{A}$ are a^F, $R^{(M)}$ and the direction cosines, n_i.

The formula (5.14) has been applied, for example, by Hutchison and McKay[144] to determine the proton coordinates in Nd(III)-doped lanthanum nicotinate dihydrate crystals, and by Balmer et al.[101] to localize the charge compensator H^+ in Co(II)-doped α-Al_2O_3.

It should be noted that there is some contradiction in (5.14) in the sense that on the one hand the unpaired electron is assumed to be entirely localized at the central ion, on the other hand the Fermi contact term a^F at the nucleus under consideration is different from zero. If a preponderant fraction of unpaired spin is localized at the central ion, both the spin-only and the orbital contributions can be described by the point-dipole approximation, so that (5.14) gives an adequate description of hf interactions of ligand nuclei. In transitional metal complexes with high covalency, i.e. with large spin densities on the ligand atoms, however, (5.14) has to be used with care. In order to be consistent in such cases, the contribution of the orbital moment at the ligand N to the $\mathbf{A}$ tensor (described by the operator $\mathcal{H}_{LI} = P\tilde{I}L/r^3$) should be included in the calculation[125, 126]. Recently, a procedure for estimating the $\mathbf{A}^{DD}$ and $\mathbf{A}^L$ tensors in copper complexes has been suggested by Atherton and Horsewill[132].

For nuclei other than protons, the one-center contribution (5.4) has to be added to (5.14). If the principal axes of $(\mathbf{A}^{DD})_1$ do not coincide with the axes of the tensor $\mathbf{T}$, the contributions arising from the term a^F, $\mathbf{T}$ and $(\mathbf{A}^{DD})_1$ cannot be individually determined from the experimental $\mathbf{A}$ tensor without using additional assumptions concerning the orientations of $\mathbf{T}$ and $(\mathbf{A}^{DD})_1$.

If the principal axes of $(\mathbf{A}^{DD})_1$, $\mathbf{T}$ and $\mathbf{g}$ coincide, the MO treatment introduced by Maki and McGarvey to interpret EPR spectra of transition metal complexes[145] is often used in ENDOR investigations. In the following, we collect some of the relevant formulae from this approach. For a σ-orbital involving a ligand atom N in the first coordination sphere, and pointing along the metal-ligand direction (principal axis g_x), the principal values of the ligand hfs tensor are given[66, 146, 147] by

$$\begin{aligned} A_1 &= a_{iso}(s) + A_{\|}(p) + A_{\|}(D) \\ A_2 &= a_{iso}(s) + A_{\perp}(p) + A_2(D) \\ A_3 &= a_{iso}(s) + A_{\perp}(p) + A_3(D) \ . \end{aligned} \tag{5.15}$$

In the expressions (5.15) the terms are defined by:

$$a_{iso}(s) = a^F ,$$

$$A_{\|}(p) = (A_1^{DD})_1 = \frac{4}{5} P\langle r_N^{-3}\rangle f_p^N$$

$$A_{\perp}(p) = (A_2^{DD})_1 = (A_3^{DD})_1 = -\frac{2}{5} P\langle r_N^{-3}\rangle f_p^N , \tag{5.15 a}$$

and

$$A_{\|}(D) = (A_1^{DD})_{2,1} + A_1^L = 2\mu_B\mu_N g_n(\varrho_M g_e + \Delta g_x)/R^{(M)3}$$

$$A_2(D) = (A_2^{DD})_{2,1} + A_2^L = -\mu_B\mu_N g_n(\varrho_M g_e + \Delta g_y)/R^{(M)3} \tag{5.15 b}$$

$$A_3(D) = (A_3^{DD})_{2,1} + A_3^L = -\mu_B\mu_N g_n(\varrho_M g_e + \Delta g_z)/R^{(M)3} .$$

In (5.15 a), f_p^N denotes the spin density of the p-orbital of ligand N. The hybridization ratio n^2 defined by the σ-orbital $\sigma = np + (1 - n^2)^{1/2}s$ is then given by $n^2 = f_p^N/\varrho_N$, where $\varrho_N = f_s^N + f_p^N$ denotes the total σ-spin density on ligand N. In the case of sp^2 hybridization, $n^2 = 2/3$. Multipole expansion of the d-orbitals[125)], core polarization[124)], and orbital reduction fractors[148)] have been neglected in (5.15).

The desription of the ligand hfs tensor outlined above is based on a perturbation approach which assumes the energy difference between the excited states and the ground state to be large compared with the matrix elements of the spin-orbit coupling operator. For low-lying excited states, the perturbation treatment is no longer valid, and the Kramers doublet base[2)] has to be applied to describe the ligand hf interactions. This more general approach has been used, for example, by Rudin et al.[12)] to interpret the hf data of Co(acacen).

5.2 Nuclear Quadrupole Interactions

The spin Hamiltonian for the nuclear quadrupole interaction has the form[2, 149, 150)]:

$$\mathcal{H}_Q = \tilde{\mathbf{I}}\mathbf{Q}\mathbf{I} \tag{5.16}$$

which writes in the principal axes system of the **Q** tensor

$$\mathcal{H}_Q = Q_1 I_1^2 + Q_2 I_2^2 + Q_3 I_3^2 \tag{5.16 a}$$

$$= Q'\left\{I_3^2 - \frac{1}{3} I (I + 1)\right\} + Q'' (I_+^2 + I_-^2) \tag{5.16 b}$$

$$= \frac{e^2 qQ}{4 I(2 I - 1)h} \{(3 I_3^2 - \mathbf{I}^2) + \eta(I_1^2 - I_2^2)\} . \tag{5.16 c}$$

In (5.16 c) $eq = eq_{33} = V_{33}$ denotes the electric field gradient (EFG) and $\eta = (V_{11} - V_{22})/V_{33}$ ($|V_{33}| > |V_{22}| > |V_{11}|$, $0 < \eta < 1$) the asymmetry parameter. Since the quadrupole tensor **Q** is traceless, it is determined, apart from its orientation, by only two parameters. Usually, in the literature, the two quantities e^2qQ/h and η are given. The relations between the coupling constants in (5.16a), (5.16b) and (5.16c) for frequently occurring nuclear spin quantum numbers are summarized in Appendix C.

The EFG at a nucleus N in a molecule is defined by

$$eq = -e \sum_j \left\langle \psi_j \left| \frac{3\cos^2\theta - 1}{r^3} \right| \psi_j \right\rangle n_j + e \sum_{k \neq N} Z_k \left(\frac{3\cos^2\theta_k - 1}{R^{(k)3}} \right) . \qquad (5.17)$$

The first term of (5.17) describes the electronic contributions to the EFG, with ψ_j referring to the occupied molecular orbitals and n_j to the number of electrons on the jth orbital. The second term denotes the contributions of the nuclei with charges Z_k to the EFG.

The crucial point in the determination of the quadrupole coupling constants is the computation of the electronic contributions which in principle should include all the core as well as the valence electrons. The computation may be carried out at various levels ranging from empirical Townes-Dailey (TD) type models[151)] to *ab initio* molecular orbital calculations[152)]. In the following section, we shall discuss two simple approaches which are often used to interpret nuclear quadrupole data of central ions and nitrogen ligands in transition metal compounds[153–156)].

5.2.1 Nuclear Quadrupole Interactions of Transition Metal Ions

In a very rough approximation, the EFG at the central ion of a transition metal complex can be traced back to two contributions, one from the valence electrons of the transition metal ion, and the other from the ligand electrons and nuclei[157)]:

$$eq = e(1 - R_d)q_{val} + e(1 - \gamma_\infty)q_{lig} \qquad (5.18)$$

with

$$eq_{val} = -e \sum_j N_j \langle 3\cos^2\theta_j - 1 \rangle \langle r_j^{-3} \rangle \qquad (5.18a)$$

and

$$eq_{lig} = -e \sum_L \sum_i N_{Li} \langle 3\cos^2\theta_{Li} - 1 \rangle \langle r_{Li}^{-3} \rangle + e \sum_L Z_L (3\cos^2\theta_L - 1)(R_L^{-3}) . \qquad (5.18b)$$

The quantity $(1 - R_d)$ represents the Sternheimer shielding factor for the valence d-electrons, and $(1 - \gamma_\infty)$ the antishielding factor for the ligand electrons. These quantities describe the polarization of the core by the EFG of the valence and ligand electrons, respectively. N_j and N_{Li}, which denote the electron populations of the jth valence orbital and of the ith orbital at the ligand nucleus L, are usually obtained from an MO calculation. Expectation values $\langle r_j^{-3} \rangle$ for free ions obtained with Hartree-Fock atomic orbitals are found in the literature[124)]. The contributions from the ligand electrons may be approximated as originating from point charges.

To improve the poor agreement between experimental and computed quadrupole coupling constants of Cu(II)[157, 158] and Co(II)[61, 159] ions, anisotropic contraction of 3d-orbitals[61, 160] or optimization of $(1 - \gamma_\infty)$[158] have been included in the calculations. The contraction of 3d-orbitals is rationalized by the fact that in a planar complex the interaction of the electrons in the in-plane orbitals ($3\,d_{x^2-y^2}$, $3\,d_{xy}$) with the negatively charged ligand atoms exceeds that of the electrons in the out-of-plane orbitals ($3\,d_{xz}$, $3\,d_{yz}$, $3\,d_{zz}$). Using this anisotropic contraction, the computed quadrupole interactions are often found to be in qualitative agreement with the experimental data[61, 160].

5.2.2 Townes-Dailey Description of the ^{14}N Nuclear Quadrupole Interaction

Most of the nitrogen quadrupole data obtained by NQR in free ligands and diamagnetic metal complexes, and by ENDOR in paramagnetic transition metal compounds, have been interpreted in terms of the TD model[150, 151]. This approach simply assumes that the field gradient at the nitrogen nucleus is due to different total populations of the three 2p-orbitals. The model may be applied to study relationships between and trends of observed quadrupole parameters of ^{14}N in series of similar free ligands and of their corresponding metal complexes, rather than to compute orbital populations[150].

The formulation of the TD theory, which is given below, follows the work of Hsieh et al.[161] and has been used in the discussion of most of the ^{14}N quadrupole data obtained from ENDOR spectra (Sect. 6).

The geometry about a ligand nitrogen coordinated to a metal ion is assumed to be planar with bond directions defined in Fig. 28. The hybridizations of the nitrogen orbitals for different angles between the bonds in the plane can be written as:

$$\begin{aligned}
\psi_{NM} &= A\,2s + B\,2p_z \\
\psi_{NK} &= C\,2s + D\,(2p_z\cos\gamma - 2p_y\sin\gamma) \\
\psi_{NL} &= E\,2s + F\,(2p_z\cos\beta + 2p_y\sin\beta) \\
\psi_\pi &= 2p_x\,,
\end{aligned} \tag{5.19}$$

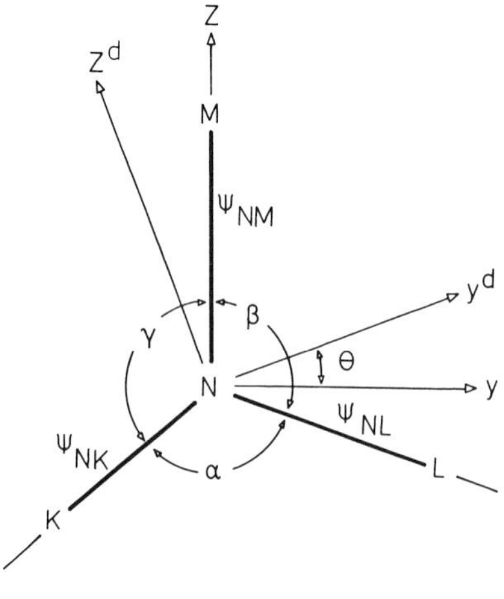

Fig. 28. Orientations of molecule-based and $\mathbf{Q}^N$ tensor axes about a planar nitrogen bound to three different groups. (Adapted from Ref. 161)

where the coefficients A, ..., F are expressions containing the bond angles α, β, γ. The non-vanishing components of the EFG tensor are

$$\begin{aligned}
q_{11}/q_0 &= -\frac{1}{2}\{B^2\ell + D^2\sigma_2 + F^2\sigma_3\} + \pi \\
q_{22}/q_0 &= -\frac{1}{2}B^2\ell + \left(-\frac{1}{2}\cos^2\gamma + \sin^2\gamma\right)D^2\sigma_2 + \left(-\frac{1}{2}\cos^2\beta + \sin^2\beta\right)F^2\sigma_3 - \frac{1}{2}\pi \\
q_{33}/q_0 &= B^2\ell + \left(\cos^2\gamma - \frac{1}{2}\sin^2\gamma\right)D^2\sigma_2 + \left(\cos^2\beta - \frac{1}{2}\sin^2\beta\right)F^2\sigma_3 - \frac{1}{2}\pi \\
q_{23}/q_0 &= \frac{3}{4}(\sin 2\beta F^2\sigma_3 - \sin 2\gamma D^2\sigma_2) \,.
\end{aligned} \tag{5.20}$$

In (5.20) ℓ, σ_2, σ_3 and π are the occupancies of the orbitals ψ_{NM}, ψ_{NK}, ψ_{NL} and ψ_π, respectively. The coefficients B^2, D^2 and F^2 are given by

$$B^2 = \frac{-\cos\alpha}{\cos\beta\cos\gamma - \cos\alpha}\,, \quad D^2 = \frac{-\cos\beta}{\cos\alpha\cos\gamma - \cos\beta}\,, \quad F^2 = \frac{-\cos\gamma}{\cos\alpha\cos\beta - \cos\gamma}\,. \tag{5.21}$$

The quantity eq_0 is the EFG due to a single electron in a nitrogen 2p-orbital. Estimates of e^2Qq_0/h range from -8.4 to -10.0 MHz[150)].

The relationship between the components of the EFG tensor in its principal axes frame, and in the coordinate system defined in Fig. 28 is given by

$$\begin{aligned}
q_{33} &= q_3^d\cos^2\theta + q_2^d\sin^2\theta \\
q_{22} &= q_2^d\cos^2\theta + q_3^d\sin^2\theta \\
q_{11} &= q_1^d \\
q_{23} &= \frac{1}{2}(q_3^d - q_2^d)\sin 2\theta \,.
\end{aligned} \tag{5.22}$$

From (5.20) and (5.22) and the expression $q_i^d = 2Q_ih/e^2Q$, which relates the principal values q_i^d of the EFG tensor to the experimentally determined components Q_i of the tensor $\mathbf{Q}$, one obtains the following formulae for the angle θ and the orbital populations:

$$\sin 2\theta = \frac{3e^2Qq_0}{4(Q_3 - Q_2)h}(\sin 2\beta F^2\sigma_3 - \sin 2\gamma D^2\sigma_2) \tag{5.23}$$

$$\ell = \left\{\frac{4(Q_3 - Q_2)h}{3e^2Qq_0}\cos 2\theta - \cos 2\gamma D^2\sigma_2 + \cos 2\beta F^2\sigma_3\right\}/B^2 \tag{5.23 a}$$

$$\pi = \frac{1}{2}(B^2\ell + D^2\sigma_2 + F^2\sigma_3) + \frac{2Q_1h}{e^2Qq_0}\,. \tag{5.23 b}$$

For a free ligand, $\beta = \gamma = \pi/2 - \alpha$, $B^2 = n^2$ and $D^2 = F^2 = (2 - n^2)/2$, where $n^2 = 1 - \cot^2(\alpha/2)$ denotes the hybridization ratio. In this case (5.23) (5.23 a) and (5.23 b) write

$$\sin 2\theta = \frac{3\,e^2Qq_0}{4(Q_3 - Q_2)h}\,(1-n^2)^{1/2}(\sigma_3 - \sigma_2) \tag{5.24}$$

$$\ell = \frac{4(Q_3 - Q_2)h}{3\,e^2Qq_0n^2}\cos 2\theta + \frac{1}{2}(\sigma_2 + \sigma_3) \tag{5.24 a}$$

$$\pi = \frac{1}{2}\left\{n^2\ell + \frac{1}{2}(2 - n^2)\,(\sigma_2 + \sigma_3)\right\} + \frac{2\,Q_1h}{e^2Qq_0}\,. \tag{5.24 b}$$

Since **Q** is traceless, the occupancies ℓ, π, σ_2 and σ_3 in the four orbitals cannot be determined from two independent values Q_i, or from e^2Qq/h and η. The number of parameters that can be calculated from the data has therefore to be reduced by suitable assumptions. In the metal-nitrogen fragment, the angle θ can be calculated from (5.23) using values of σ_2 and σ_3 which are reasonable from a chemical point of view (i.e., they should reflect the relative electronegativities of nitrogen and of the coordinated atoms in the ligand). The occupancy ℓ of the orbital ψ_{NM} can be determined from (5.23 a) with estimated values for σ_2 and σ_3. In the free ligand the occupancy ℓ of the lone pair orbital is usually assumed to be 2. Upon metal complexation this value is reduced by nitrogen-metal σ-electron donation. Finally, the occupancy π of the ψ_π orbital is obtained using (5.23 b).

6 Discussion of the Literature

This section covers the literature of ENDOR on transition metal complexes with organic ligands through the end of 1980. The discussion also includes unpublished results and papers submitted for publication which came to our knowledge. Only contributions for which ENDOR spectroscopy yields either new or significantly improved results compared with the data obtained from EPR measurements will be discussed in more detail.

6.1 Cu(II) and Ag(II) Compounds (d^9)

Most ENDOR investigations on transition metal complexes with organic ligands have been done on Cu(II) compounds. Single crystal EPR spectra of copper complexes are normally characterized by hf structures of the copper ion (I = 3/2) and the ligand nuclei in the first coordination sphere[162, 163]. The spectra are often complex due to overlapping resonances of the isotopes ^{63}Cu and ^{65}Cu and partly resolved proton hfs. Thus, the information obtained from EPR spectra concerning hf interactions is often quite restricted. Moreover, the intricate ligand hf structure usually prevents the determination of the quadrupole tensor of the Cu nucleus from second order line shifts or from "forbidden" EPR transitions.

In the corresponding ENDOR spectra, the hf and quadrupole splittings of central ion and ligand nuclei are, in general, fully resolved. The presence of two copper isotopes does not significantly influence the width of the ligand ENDOR lines.

In copper and silver complexes, optimum ENDOR enhancements for all types of nuclei are usually achieved for tempeatures between 10 to 30 K. For Cu-ENDOR in copper containing proteins, temperatures $\leqslant 4$ K are sometimes required (see below).

6.1.1 Planar Cu(II) and Ag(II) Complexes

Cu(II)-bis(oxychinolate), $Cu(Ox)_2$

The first extended ENDOR study on a transition metal complex with organic ligands was reported by Rist and Hyde[15, 75]. The authors applied ENDOR on $Cu(Ox)_2$ substituted into single crystals of phthalimide (PHI) and 8-hydroxyquinoline (HOx) and determined the hfs and nuclear quadrupole tensors of the two magnetically equivalent nitrogens as well as the hfs tensor of the protons in position 2 (see Fig. 29). The magnetic parameters obtained from the angular dependence of the ENDOR spectra are collected in Table 3. The nitrogen hf coupling constants of $Cu(Ox)_2$ are considerably weaker in the HOx than in the PHI lattice. This indicates that the $\mathbf{A}^N$ tensor is sensitive to changes in the packing of the metal complex. In contrast to this, the nitrogen quadrupole tensor $\mathbf{Q}^N$ is not remarkably influenced by the lattice forces.

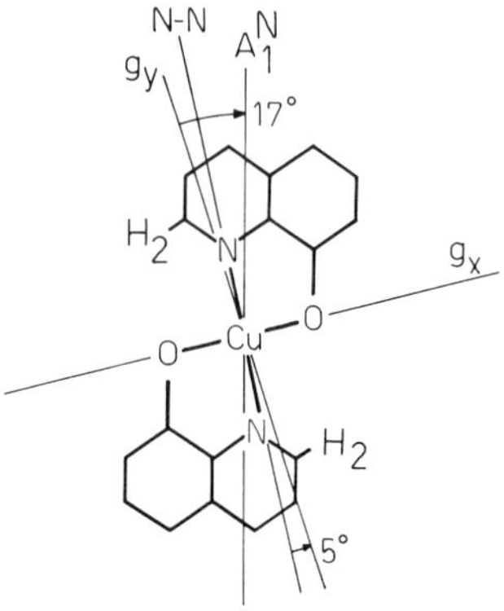

Fig. 29. Directions of **g** and $\mathbf{A}^N$ principal axes in $Cu(Ox)_2$. (Adapted from Ref. 15)

Table 3. ENDOR parameters of $Cu(Ox)_2$ (data from Rist and Hyde[15]; in MHz)

Tensor		Host lattice	
		PHI	HOx
$\mathbf{A}^N$	A_1	40.3	33.4
	A_2	29.7	28.4
	$A_3{}^a$	30.3	26.1
$\mathbf{Q}^{N\,b}$	e^2qQ/h	−2.62	−2.69
	η	<0.1	<0.1
$\mathbf{A}^{H2\,d}$	$A_1{}^c$	8.80	8.48
	A_2	3.78	2.86
	$A_3{}^a$	2.76	2.20
	a_{iso}	5.12	4.52

a Normal to the complex plane
b Largest principal axis along A_1^N
c In the complex plane
d Proton in position 2 (see Fig. 29)

The coupling tensor $\mathbf{A}^{H2}$ has been used to determine the orientation of the **g** and $\mathbf{A}^N$ tensor axes in the complex plane. The angle between the Cu–H direction and the largest principal axis of $\mathbf{A}^{H2}$ has been calculated by considering different spin densities in the ligand orbitals (4% on the nitrogen and oxygen ligands, 0.5% on the carbon atoms bonded to the protons in position 2). As illustrated in Fig. 29, the axis of the in-plane value g_x lies along the O–Cu–O bond, whereas the axis of g_y forms an angle of 5° to the Cu–N bond direction. The largest quadrupole coupling coincides with the direction of the nitrogen hf coupling A_1^N which bisects the C–N–C angle of the pyridine ring ($\sphericalangle (g_y, A_1^N) = 17°$).

The TD approach outlined in Sect. 5.2.2 has been applied to interpret the nuclear quadrupole tensor, $\mathbf{Q}^N$. Since $\mathbf{Q}^N$ is nearly axial, the π population has been approximated by $\pi \simeq (\sigma_2 + \sigma_3)/2 = \sigma$. Assuming $\sigma = 1.2$ and $e^2q_0Q/h = -8$ MHz, the occupancy of the lone pair nitrogen orbital was found to be $\ell = 1.6$. The quadrupole coupling is

Table 4. Nitrogen coupling constants A_3^N and Q_3^N for planar copper and silver complexes (data from Rist and Hyde[16]; in MHz)

Metal complex	Host	A_3^N	Q_3^N	e^2qQ/h [a]
$Cu(Ox)_2$	PHI	30.3	0.65	2.6
	$Zn(Ox)_2 \cdot 2H_2O$	27.5	0.89	3.6
$Cu(pic)_2$	PHA	36.9	0.50	2.0
	$Zn(pic)_2 \cdot 4H_2O$	29.7	0.90	3.6
$Cu(Qn)_2$	$Zn(Qn)_2 \cdot XH_2O$	29.0	0.63	2.5
$CuMe(pic)_2$	$ZnMe(pic)_2$	35.1	0.58	2.3
$Cu(sal)_2$	$Pd(sal)_2$	43.8	≤ 0.13	
$Cu(dim)_2$	$Ni(dim)_2$	47.2	0.38	
$Ag(pic)_2$	$Zn(pic)_2 \cdot 4H_2O$	50.1	0.8	
	PHA	~51		

[a] Axial symmetry is assumed

slightly larger in the HOx than in the PHI host, thus, ℓ[HOx] > ℓ[PHI]. This is consistent with the stronger Cu–N bond observed in $Cu(Ox)_2$: PHI ($a_{iso}^N(PHI) > a_{iso}^N(HOx)$).

In a subsequent paper Rist and Hyde[16] introduced the technique of orientation selection in powder samples (Sect. 4.1) and applied this method to a series of planar copper and silver complexes. The data obtained from single crystal-like ENDOR spectra with $\mathbf{B}_0$ oriented along the complex normal are collected in Table 4. In all compounds listed in this table, the spin density of the ligands is positive and resides in σ-orbitals, in accordance with the ground state of the metal ions. Also, the hf coupling of the silver complex is much larger than the one of the corresponding copper complex in the same host lattice, revealing that the metal-ligand interaction is stronger in the $4d^9$ than in the corresponding $3d^9$ compound.

Coordination of water molecules at the axial positions of copper compounds ($Cu(Ox)_2$ in $Zn(Ox)_2 \cdot 2H_2O$ and $Cu(pic)_2$ in $Zn(pic)_2 \cdot 4H_2O$) weakens the metal-ligand interactions in the complex plane. This trend is also reflected in an increase $\Delta\ell$ of the lone pair orbital population ℓ, which amounts to $\Delta\ell[Cu(Ox)_2] = \ell[Zn(Ox)_2 \cdot 2H_2O] - \ell[PHI] = 0.12$ and $\Delta\ell[Cu(pic)_2] = \ell[Zn(pic)_2 \cdot 4H_2O] - \ell[PHA] = 0.20$ electrons, respectively.

$Cu(sal)_2$ (see below) shows a surprisingly strong nitrogen hf interaction which is about one third larger than common nitrogen coupling constants of copper complexes with mixed nitrogen-oxygen ligands. This strong hf coupling might be traced back to the interaction of the two salicylaldoxime ligands through intramolecular hydrogen bonds.

Cu(II)-bis(salicylaldoxime), $Cu(sal)_2$

An EPR and ENDOR investigation of the planar copper complex $^{63}Cu(sal)_2$ (Fig. 30) substituted into a single crystal of $Ni(sal)_2$ has been reported by Schweiger et al.[62, 65]. The aim of this work was to determine the structure of the internal H-bond occuring in $Cu(sal)_2$, and to draw a detailed picture of the unpaired electron distribution on the ligands. ^{1}H-, ^{14}N- and ^{63}Cu-ENDOR spectra appeared in distinct rf ranges and were observed for numerous crystal settings. Proton ENDOR spectra of two simultaneously excited overlapping sites could be separated by choosing B_0 approximately equal to a resonance field corresponding to the maximum of an EPR line of one site, coincident with the minimum of an EPR line of the second site. Thus, the ENDOR signals of the two sites differ in phase by $\simeq 180°$, as illustrated in Fig. 31.

Because of the inversion-symmetry of the complex, all ligand nuclei are magnetically equivalent in pairs. The typical eight-line spectrum for two magnetically equivalent $I = 1$ nuclei (Fig. 9) and the extra splitting of these transitions for $\overline{Q}_{33}(m_S) \simeq 0$ (Fig. 11) was first observed in nitrogen ENDOR spectra of $Cu(sal)_2$[62,65]. The ENDOR data in Table 5.1 (nitrogens and four pairs of protons) were evaluated from the rotation patterns using the perturbation approach outlined in Sect. 3.

Proton ENDOR. For the assignment of the proton hfs tensors to corresponding protons in the $Cu(sal)_2$ molecule (Fig. 30) or in the $Ni(sal)_2$ host, the point-dipole formula (5.6) was applied. By symmetry arguments, one of the principal axes of the coupling tensors of the protons belonging to the planar $Cu(sal)_2$ compound has to lie *normal* to the complex plane. Indeed, the coupling tensors of H 11, H 15 and H 16 are found to behave in this

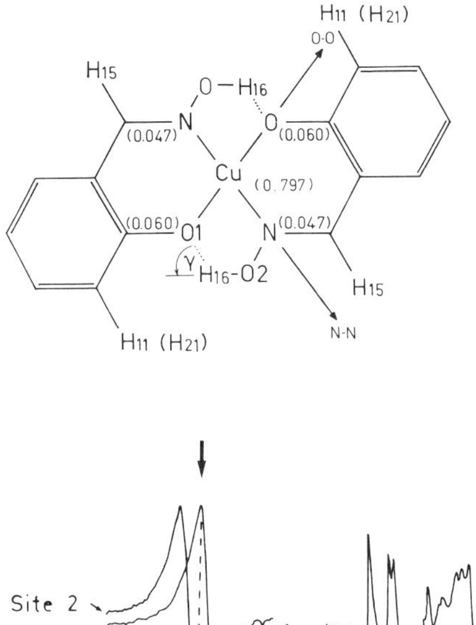

Fig. 30. Structure of $Cu(sal)_2$. Numbers in parentheses denote Mulliken populations. (From Ref. 62)

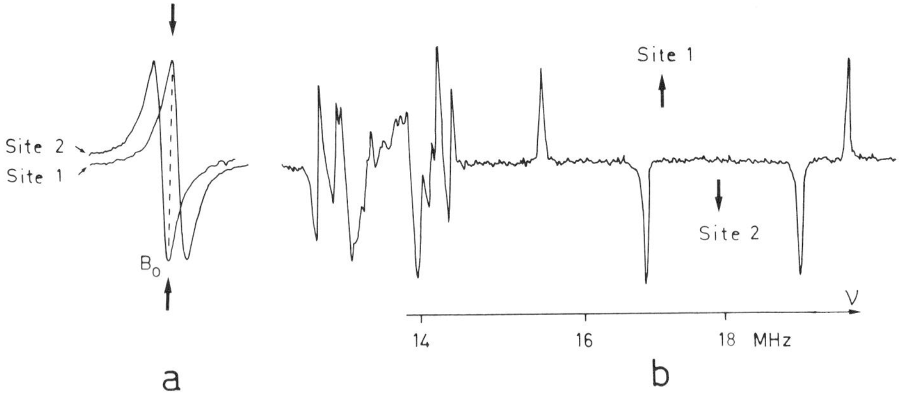

Fig. 31 a, b. Proton ENDOR spectrum of two overlapping sites of $Cu(sal)_2$. **a)** Arrow indicates the choice of the $\mathbf{B}_0$ field position in the EPR spectrum. **b)** Differentiation of ENDOR signals of overlapping sites by means of the phases of their signals. (Adapted from Ref. 62)

Table 5.1. Magnetic parameters of $Cu(sal)_2$ in $Ni(sal)_2$ (data from Schweiger and Günthard[62]; A_i and Q_i in MHz)

EPR data

	Principal values x	y	z
$\mathbf{g}$[a]	2.056	2.039	2.203
$\mathbf{A}^{Cu}$[b]	-109	-111	-609

ENDOR data[c]

	Principal values 1	2	3[d]		φ_0[e]	d(Cu–H) Point-dipole coupling	Crystal[f] structure
$\mathbf{A}^{N}$	51.96	42.10	43.64	(43.8)[g]	7		
$\mathbf{Q}^{N}$[h]	-1.71	1.91	-0.2		21		
$\mathbf{A}^{H11}$	-1.26	1.78	-1.89		15	4.29	4.52
$\mathbf{A}^{H15}$	13.00	9.15	8.48	(8.58)[g]	20	3.84	3.87
$\mathbf{A}^{H16}$	6.60	-0.87	-5.97	(5.62)[g]	-45	2.89	
$\mathbf{A}^{H21}$	-1.95	-1.08	3.19			3.71	3.84

[a] g_x includes an angle of $-3°$ with the Cu–N bond direction (in the positive sense, Fig. 30)
[b] Coaxial to $\mathbf{g}$
[c] For direction cosines, see[62]
[d] Principal axis of A_3 (Q_3) normal to the complex plane, parallel g_z (except for H21)
[e] Angle between the principal axis of A_1 (Q_1) and the Cu–N bond direction
[f] From Ref. 164
[g] Powder data of $Cu(sal)_2$ in $Pd(sal)_2$, from Ref. 16
[h] $e^2qQ/h = 2Q_2 = 3.82$ MHz, $\eta = 0.79$

Table 5.2. Bonding parameters in $Cu(sal)_2$[62]

f_s^N	0.029	U_{Cu}[a]	0.797
f_p^N	0.054	U_N	0.095
ϱ_N	0.083	(U_O)	(0.120)
n_N^2[b]	0.651		

[a] Mulliken populations[129]
[b] n_N^2 defined by $f_p^N/(f_s^N + f_p^N)$[62]

manner. Since none of the principal axes of H21 is normal to the complex plane, this proton belongs to one of the $Ni(sal)_2$ molecules surrounding the Cu-complex (its corresponding position is H11 in $Cu(sal)_2$). The Cu–H distances for H11, H15 and H21 calculated from the largest principal values of the traceless dipolar tensors are in good agreement with the crystal structure[164] (Table 5.1).

No information about the structure of the fragment O1–H16–O2 is available from X-ray data. In particular, it is not known whether this structure is linear, bent, symmetric or asymmetric, and whether or not H16 lies in the complex plane. From $d(O1–O2) = 2.58$ Å, one would predict the H-bond O1–H16–O2 to have an asymmetric, bent structure[165]. This leads to two alternatives, namely min$d$(O–H16) = d(O1–H16) or d(O2–H16) (with min$d$(O–H16) = 1.03 Å[166]). The hfs tensor $\mathbf{A}^{H16}$ allows the determination of the correct alternative on the basis of the corresponding angles, γ (Fig. 30). In addition $\mathbf{A}^{H16}$ also fixes the orientation of the $\mathbf{g}$ tensor with respect to the $Cu(sal)_2$ molecule[62, 167]. Since one of the principal axes of $\mathbf{A}^{H16}$ and $\mathbf{A}^{N}$ is perpendicular to the complex plane (see above), the structure of the first coordination sphere including H16 is considered to be planar. A comparison of the observed $\mathbf{A}^{H16}$ tensor (which is nearly traceless but strongly rhombically distorted) with that predicted from a seven-center point-dipole model (including spin populations on Cu, N, O1 and O2) shows that the structure of the fragment O1–H16–O2 may be described by min$d$(O–H16) = d(O2–H16) with an upper limit of 50° for the angle γ. For this value of γ the distance d(Cu–H16) is calculated to 2.76 Å compared to 2.89 Å from the single-centre point-dipole model. The alternative condition mentioned above, namely min$d$(O–H16) = d(O1–H16), leads to implausible values for γ.

Nitrogen ENDOR. The nitrogen hfs tensor is nearly axially symmetric with the largest principal axis oriented approximately along the Cu–N bond direction ($\sphericalangle$(Cu–N, A_1^N) $\simeq$ 7°). A very similar result has been reported by Moores and Belford[168] for the planar copper complex $Cu(msal)_2$.

The experimental ^{14}N hf parameters have been analyzed within the MO framework outlined in Sect. 5.1. To get an *estimate* of the spin density at the oxygen ligands, O1, which can be utilized in the multi-center point-dipole model mentioned above, the oxygen (^{17}O) spin density data have been taken from $Cu(pic)_2$ in $Zn(pic)_2 \cdot 4H_2O$[147] by assuming (A_i^N/A_i^O) $[Cu(sal)_2] \simeq (A_i^N/A_i^O)$ $[Cu(pic)_2]$. Some of the relevant results of the analysis of the ligand hfs tensor are given in Table 5.2 and Fig. 30.

The nitrogen quadrupole parameters (Table 5.1) have recently been analyzed by Murgich[169] in terms of the TD theory (Sect. 5.2.2). The principal axis of the largest coupling constant (Q_2^N) approximately bisects the Cu–N–C angle. This orientation of the $\mathbf{Q}^N$ tensor axes corresponds to the "anomalous axes" predicted by Sauer and Bray[170] for some unsymmetrically substituted aldoximes. The populations of the 2p-orbitals in $Cu(sal)_2$ are compared with the corresponding populations in the free Sal ligand. A charge transfer of about $\Delta\ell = 0.35$ from the nitrogen lone-pair orbital to the central ion upon metal complexation has been found[169].

Cu(II) and Ag(II) tetraphenylporphyrin, Cu(TPP), Ag(TPP)

An extensive EPR and ENDOR study of Cu(TPP) and Ag(TPP) (Fig. 32) doped into $(H_2O)Zn(TPP)$ single crystals has been published by Brown and Hoffman[66, 171]. The complete set of hfs and quadrupole tensors for the ^{14}N nuclei and the hfs tensors of the Ag(II) and Cu(II) ions and of the pyrrole protons are reported. A detailed analysis of the magnetic parameters (Table 6.1), which are measured with high accuracy, has been presented using the standard MO treatment[66].

Proton ENDOR. In both metal complexes four pairs of magnetically equivalent pyrrole protons along with some weakly coupled phenyl protons have been observed.

This indicates that the paramagnetic guest compounds exhibit the center of symmetry required by the Zn(TPP) host crystal structure. Two types of pyrrole protons (1, 3 and 2, 4, Fig. 32) with slightly different hf values are found in Ag(TPP), indicating that the site symmetry of Ag(TPP) cannot be higher than S_4 or C_4.

In the corresponding copper complex all eight protons are geometrically equivalent within error limits. The metal-proton distances in Ag(TPP) calculated by the classical

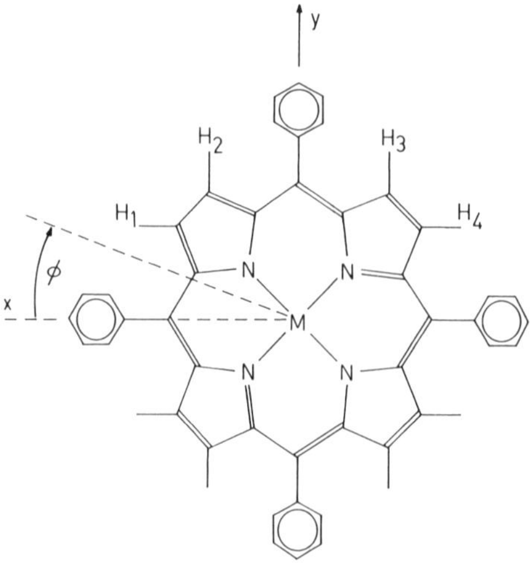

Fig. 32. Structure of *M*(TPP). (Adapted from Ref. 66)

Table 6.1. Magnetic parameters of Ag(TPP) and Cu(TPP) in $(H_2O)Zn(TPP)$ (data from Brown and Hoffman[66]; in MHz)

M	Atom	A_1	A_2	A_3	Q_1	Q_2	Q_3	φ_0[a]
Ag	^{107}Ag	174[b]	88.376	88.376				
	^{109}Ag	187[b]	102.001	102.001				
	^{14}N	78.870	61.330	62.918	−0.692	0.915	−0.222[c]	45
	H1	3.566	1.729	1.232				24.7
	H2	3.443	1.618	1.145				65.1
	H3	3.578	1.749	1.232				114.6
	H4	3.462	1.648	1.145				155.4
Cu	^{63}Cu	−615[b]	−102.7	−102.7				
	^{14}N	54.213 (60.3)	42.778 (47.7)	44.065 (47.7)[d]	−0.619	0.926	−0.307[c]	45
	H1	2.52	0.74	0.80				27.5
	H2	2.45	0.71	0.80				64.0
	H3	2.41	0.66	0.80				117.2
	H4	2.49	0.69	0.80				154.3

[a] Orientation of the largest principle axis (Fig. 32) in the complex plane
[b] Obtained from EPR spectra
[c] Obtained from the requirement $\Sigma Q_i = 0$
[d] From an $X\alpha$ calculation[172]

Table 6.2. Mulliken populations and hybridization ratios[66)]

	Ag	Cu
U_M	0.379	0.624
U_N	0.155	0.099
$n_{ex}{}^{b}$	2.38	1.85(3.4)^c
$n_{120}{}^{d}$	2.00	2.00
n_{geom}	0.94	0.88

[a] Contribution of the N_{1s} orbital is included
[b] n_{ex} = p/s ratio = $n^2/(1 - n^2)$, see Sect. 5.1
[c] From an $X\alpha$ calculation[172)]
[d] $n_{120} = \tan^2 120° - 1$
[e] $n_{geom} = \tan^2\alpha - 1$

point-dipole approach (with the unpaired electron entirely localized at the metal ion) are about 0.4 Å less than those determined from X-ray measurements. Moreover, the dipolar part of the proton hfs tensors are strongly rhombically distorted. These properties of the tensors indicate that there is substantial delocalization of unpaired spin over the porphyrin ring in Ag(TPP).

Considering a five-point spin density distribution (central ion and four nitrogens) for the determination of the dipolar proton hfs tensors in Ag(TPP) (5.5), the computed $\mathbf{A}^{DD}$ principal values are found to be close to the experimental results. It should be noted that in Ag(TPP) the Mulliken population, U_N, on the nitrogen nearest to the pyrrole proton provides a *larger* contribution to $\mathbf{A}^{DD}$ along the Ag-H direction than the population U_{Ag}.

As a consequence of the smaller covalence of Cu(TPP), the pyrrole proton tensors are nearly axially symmetric and the Cu-H distances calculated with the entire unpaired electron at the Cu(II) ion are in excellent agreement with X-ray data. The difference in covalency of Ag(TPP) and Cu(TPP) is also reflected by the s-spin densities on the pyrrole protons which amount to $\varrho_H(Ag) = 0.15\%$ and $\varrho_H(Cu) = 0.093\%$, respectively. A comparison with the corresponding data of an $X\alpha$ calculation on Cu(II)-porphine[172)] ($\varrho_H(Cu) = 0.071\%$) indicates that the state-of-the-art electronic structure calculation *underestimates* the amount of unpaired spin density on the porphyrin ring.

Nitrogen ENDOR. For general orientations of $\mathbf{B}_0$, the two pairs of inversion-related, and thus magnetically equivalent, nitrogens in Cu(TPP) and Ag(TPP) show ENDOR patterns which are well described by the second order expressions (B 5) given in Appendix B. More complex spectra are observed when $\mathbf{B}_0$ lies in one of the mirror planes S_{xz} or S_{yz}. This is due to noncrossing effects of energy levels[67)] discussed in Sect. 3.2.2. For $\mathbf{B}_0$ oriented normal to the complex plane, a numerical diagonalization of the spin Hamiltonian matrix (including all four nitrogens) shows[68)] that the ^{14}N-ENDOR spectrum is indeed composed of a very large number (> 60) of transitions with remarkably different intensities (Fig. 10). For general $\mathbf{B}_0$ orientations, Brown and Hoffman[66)] treated each pair of the magnetically equivalent nitrogens separately by numerical diagonalization of the spin Hamiltonian matrix. Noncrossing regions near the symmetry planes, where the presence of all four nitrogens becomes relevant, have been omitted in the analysis of the data.

The ^{14}N hf coupling constants in Table 6.1 again indicate that the delocalization of unpaired spin is significantly higher in Ag(TPP) than in Cu(TPP). In both complexes the largest coupling (A_1^N) lies along the M–N bond. The hfs tensors are slightly rhombically distorted with $A_3^N (> A_2^N)$ along the complex normal. A comparable rhombic character of $\mathbf{A}^N$ tensors is reported in EPR and ENDOR studies on $Cu(msal)_2$[168], $Cu(sal)_2$[62] and $Cu(Ox)_2$[15].

The nitrogen quadrupole tensors of both *M*(TPP) complexes are found to be very similar. The largest (positive) quadrupole coupling lies in the porphyrin plane, *normal* to the M-N bond ("anomalous axes"[170]), as found in $Cu(sal)_2$[62] and Cu(salen)[173]. For the analysis of the nitrogen hf parameters, the MO theory outlined in Sect. 5.1 has been applied. Some of the relevant data obtained with this treatment are summarized in Table 6.2. The Mulliken populations $U_{Ag} = 0.38$ and $U_{Cu} = 0.62$ indicate that the M-N bond is strongly covalent. The experimentally determined hybridization ratios n_{ex} are close to the value $n_{120} = 2$. The corresponding ratios n_{geom}[5], calculated by using angles, $\alpha = \sphericalangle(CNC)$, taken from the molecular structures of the two complexes, are less than 50% of n_{120}. It has been found that n_{ex} decreases noticeably when π-bonding is included. The ^{14}N hf coupling constants (Table 6.1) and the hybridization ratio calculated for Cu(II)-porphine by the $X\alpha$ method[172] are considerably higher than the experimental values in Cu(TPP). These deviations are due to underestimated contributions of the populations N_{2s} and N_{2p} to the $\psi_{B_{2g}}$ molecular orbital. The significant rhombicity of the nitrogen hfs tensors ($A_3^N/A_2^N(Ag) = 1.026$, $A_3^N/A_2^N(Cu) = 1.0301$) is traced back to σ-π mixing. According to Brown and Hoffman[66] the ordinary ligand-field MO analysis, however, does not seem to be adequate to describe the spin density in the π-orbitals.

Nitrogen 2p-orbital populations have been calculated from the experimental quadrupole tensors using the TD theory. The lone pair populations $\ell(Ag) = 1.838$ and $\ell(Cu) = 1.901$ have been determined from the nitrogen MO coefficients (obtained from the ^{14}N hfs tensors). The populations $\sigma_2 = \sigma_3 = \sigma$ and π are calculated with both values n_{ex} and n_{geom}. They do not change substantially as the hybridization ratios are varied. With $n_{ex}(Ag) = 2.38$ and $n_{ex}(Cu) = 1.85$, populations $\sigma(Ag) = 1.500$, $\sigma(Cu) = 1.548$ and $\pi(Ag) = 1.667$, $\pi(Cu) = 1.731$ are obtained. It has been mentioned[66] that although the delocalization of the unpaired electron is much greater in Ag(TPP) than in Cu(TPP), the total nitrogen L-shell population N_T is only 3% greater for Cu(TPP) ($N_T(Ag) = 6.505$, $N_T(Cu) = 6.728$).

Metal ENDOR. ENDOR spectra from the central ions ^{63}Cu, ^{107}Ag and ^{109}Ag have been recorded for $\mathbf{B}_0$ in the porphyrin plane (Table 6.1). In both compounds the metal hfs tensors are axially symmetric within experimental error. The ratio of the $A_\perp^{Ag}$ values of the two Ag isotopes, $A_\perp(^{107}Ag)/A_\perp(^{109}Ag) = 0.8664 \pm 0.0002$, is significantly smaller than the ratio of the two nuclear g-factors $g_n(^{107}Ag)/g_n(^{109}Ag) = 0.8698$. This difference arises from the penetration of unpaired s-electron density into the nuclei ^{107}Ag and ^{109}Ag which have different nuclear charge distributions (hyperfine anomaly[2]).

---

5 $n_{ex} = n^2/(1 - n^2)$, where n^2 is defined in Sect. 5.1; $n_{geom} = \tan^2\alpha - 1$, $n_{120} = n_{geom}$ ($\alpha = 120°$)

[N,N'-ethylene-bis(salicylideniminato)]Cu(II), Cu(salen)

A single crystal nitrogen and proton ENDOR study of Cu(salen) (Fig. 33) doped into Ni(salen) has been published by Kita et al.[173, 174]. The magnetic parameters obtained for this *Schiff* base complex are summarized in Table 7.

Proton ENDOR. Cu-H directions and distances obtained from ENDOR data for various protons are found to be in good agreement with the crystal structure. The isotropic proton hfs in the ethylenediamine bridge are considerably larger for the equatorial protons than for the axial protons. An opposite behaviour has been found in Co(acacen)[12] (see below), reflecting the different ground states of this type of planar Cu(II) and Co(II)*Schiff* base complexes. The protons at positions 7 and 7′ are characterized by large isotropic hfs constants of about 20 MHz. Similar coupling constants have been found for the corresponding protons in a number of planar Cu(II) complexes (a^H_{iso}[Cu(sal)$_2$][62] < a^H_{iso}[Cu(msal)$_2$][168] < a^H_{iso}[Cu(salim)$_2$][175] < a^H_{iso}[Cu(amben)][176] < a^H_{iso}[Cu(salen)][173]).

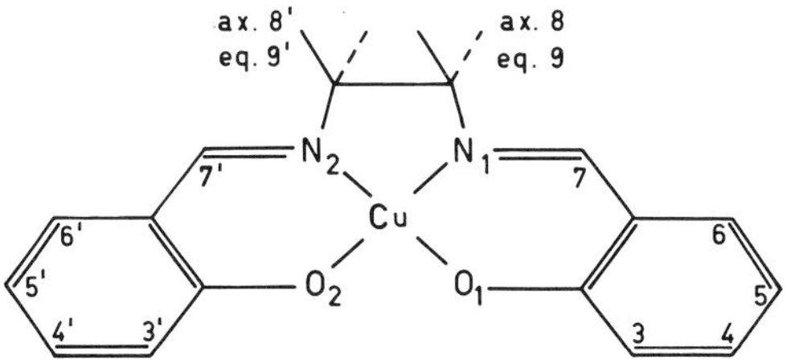

Fig. 33. Structure of Cu(salen). (Adapted from Ref. 173)

Table 7. Proton and nitrogen parameters of Cu(salen) in Ni(salen) (data from Kita et al.[173]; in MHz)

Hyperfine coupling parameters of Cu(salen) protons[a]				
Position	A_1	A_2	A_3	a_{iso}
7	22.80	19.38	18.43	20.22
7′	23.62	19.45	18.64	20.57
3	1.77	−1.34	−1.87	−0.48
3′	1.84	−1.26	−1.39	−0.27
8	5.13	−0.90	−3.47	0.76
8′	5.54	−1.41	−1.43	0.90
9	8.74	4.29	3.56	5.53
9′	9.57	5.32	4.40	6.43

Nitrogen hf and quadrupole coupling parameters[a]			
N_1	$A_1 = 50.2$	$Q_1 = -1.04$	
	$A_2 = 37.2$	$Q_2 = 0.69$	$e^2qQ/h = -2.1$
	$A_3 = 39.1$	$Q_3 = 0.36$	$\eta = 0.32$
N_2	$A_1 = 50.9$	$Q_1 = -1.21$	
	$A_2 = 37.0$	$Q_2 = 0.68$	$e^2qQ/h = -2.4$
	$A_3 = 38.8$	$Q_3 = 0.52$	$\eta = 0.13$

[a] For direction cosines, see[173]

Nitrogen ENDOR. The hfs and quadrupole tensors of the two nitrogen nuclei slightly deviate from axial symmetry. Their largest principal axes are approximately directed to the copper ion. Unpaired electron populations calculated by using a formalism similar to the one outlined in Sect. 5.1 are found to $f_s^N = 0.027$ and $f_p^N = 0.071$ for N1 and $f_s^N = 0.027$ and $f_p^N = 0.077$ for N2. For various copper complexes, the isotropic hf coupling constants of the protons at positions 7 and 7′ are found to correlate with the unpaired electron population on the nitrogen atoms, i.e. the larger f_p^N, the larger a_{iso}^{H7} and the smaller f_s^N/f_p^N the larger a_{iso}^{H7}. Comparison with spin distributions in related Cu(II) compounds further indicates that the unpaired electron delocalization on nitrogen becomes larger when the N-Cu-N arrangement changes from a *trans*- to a *cis*-configuration. Such a variation in the spin distribution has also been found in an MO model calculation[173)].

The ^{14}N nuclear quadrupole tensors have been analyzed using the TD model. Assuming $\sigma_2 = \sigma_3 = 1.20$, the occupancies ℓ and π and the net charges on the nitrogen atoms are calculated to 1.62, 1.27 and −0.29 for N1 and to 1.61, 1.23 and −0.24 for N2, respectively.

Single crystal-like ENDOR spectra ($\mathbf{B}_0$ along g_z) of Cu(salen) and some related compounds[177)] (Cu(3MeOsalen), Cu(4Mesalen) and Cu(5MeOsalen)) in frozen solutions of THF + pyridine confirm the assignment of the ENDOR transitions to the corresponding protons reported in the single crystal work discussed above. Moreover, the observed *decrease* in hfs of the chelate nitrogens and of the protons 7 and 7′ upon axial coordination of pyridine indicates that a fifth ligand causes a significant reduction of the covalent bonding in the complex plane of Cu(salen).

Preliminary Cu-ENDOR results on Cu(salen) show that the nuclear quadrupole coupling constant of the copper ions lies in the range between those for square planar O_4 and S_2O_2 coordinated copper complexes[177)].

Cu(II)-bis(N,N′-diethyl-dithiocarbamate), $Cu(et_2dtc)_2$

Proton ENDOR spectra of $Cu(et_2dtc)_2$ diluted into $Ni(et_2dtc)_2$ and $Zn(et_2dtc)_2$ single crystals have been reported by Snaathorst et al.[57)] and by Kirmse et al.[178)]. In the Ni host[57)], hfs data have been evaluated for the four methylene protons in the copper complex and for two methylene protons of the nickel compound (20B, 21B, Fig. 34). Although the number of ENDOR transitions is reduced by deuteration of the methyl groups in the ethyl substituents, the rotation patterns remained rather complex and had to be recorded in steps of 2° increments. Nuclear dipole-dipole couplings observed in the proton ENDOR spectra for certain $\mathbf{B}_0$ field orientations proved to be helpful for the assignment of the measured hfs tensors to the pair of protons in the same methylene group.

The dipolar parts of the analyzed hfs tensors have been compared with calculated values obtained from first order expressions of the electron-nuclear dipole interaction (5.3)[57)]. The coefficients of the atomic orbitals used in this computation, which considers all two- and three-center contributions, are obtained from an extended Hückel calculation (ethyl groups replaced by protons). It has been found that almost 100% of the unpaired electron is located on the CuS_4 fragment so that the replacement of the ethyl groups by protons is of minor importance for the calculation of the atomic orbital coefficients. The experimental and theoretical hfs data, summarized in Table 8, are found to

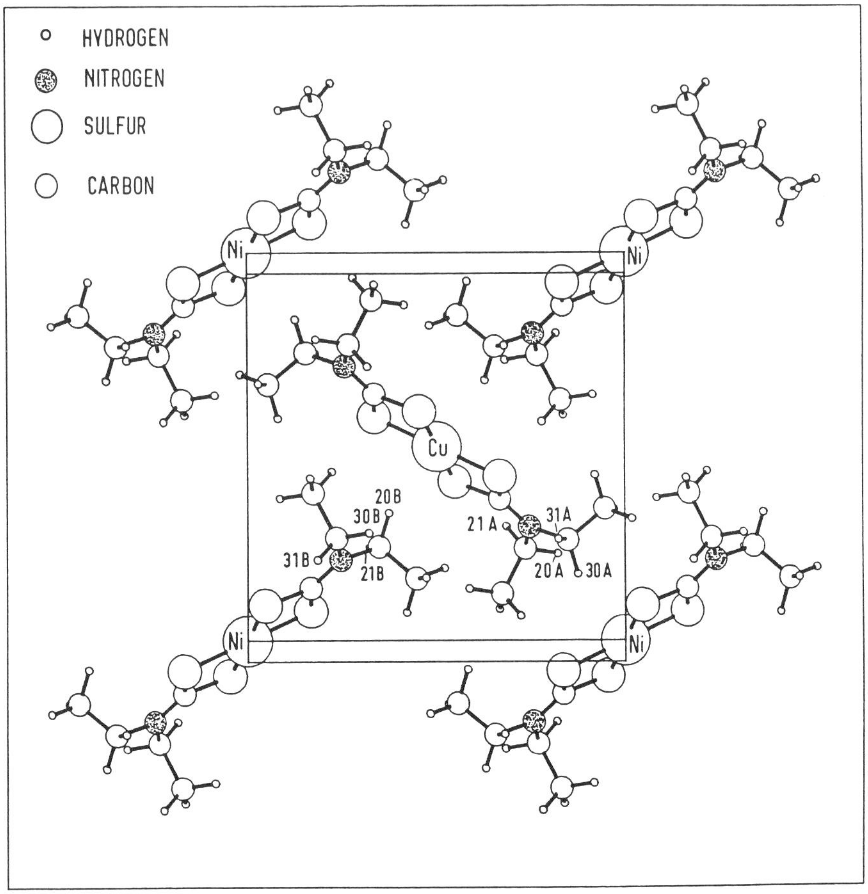

Fig. 34. Structure of $Ni(et_2dtc)_2$ (Projection in the *bc* plane). (From Ref. 57)

agree surprisingly well. Because of the high degree of covalency in the $Cu(et_2dtc)_2$ molecule, the three-center contributions range from 20 to 33% of the two-centre values and should therefore be taken into account in the calculation of the dipolar coupling constants. The directions of the proton principal axes, however, are not strongly influenced (less than 4 degrees) by the three-center contributions.

In Cu(II)-doped $Zn(et_2dtc)_2$ crystals the hf coupling constants of two protons have been reported (axial, traceless tensors with $A_{\parallel}^{H82} = 3.97$ MHz and $A_{\parallel}^{H22} = 3.49$ MHz)[178]. In the pure crystals, $Zn(et_2dtc)_2$ and $Cu(et_2dtc)_2$ form dimers of different structure[179, 180]. From comparison with both the copper and the zinc crystal structure data, the hfs tensors have been assigned to a methylene proton (H 82) in the zinc complex which forms a dimer with the copper complex and to a methylene proton (H 22) placed near the sixth coordination position of the copper atom, respectively. The positions of these two protons determined by ENDOR suggest that the most significant differences between the structures of the $(Zn(et_2dtc)_2)_2$ and the $Zn(et_2dtc)_2$-$Cu(et_2dtc)_2$ dimers are restricted to the Zn_2S_8 and $ZnCuS_8$ units.

Table 8. Proton hf coupling constants of $Cu(et_2dtc)_2$ in $Ni(et_2dtc)_2$ (data from Snaathorst et al.[57]; in MHz)

Proton[a]		Calculated[b] 2-centee	3-center	Total	Experimental[b, c]
20 A	A_1	1.20	−0.33	0.90	0.90
	A_2	−0.54	0.15	−0.39	−0.45
	A_3	−0.66	0.18	−0.51	−0.45
	a_{iso}				1.89
21 A	A_1	3.15	−0.63	2.52	2.40
	A_2	−1.44	0.30	−1.14	−0.75
	A_3	−1.71	0.33	−1.38	−1.65
	a_{iso}				0.81
30 A	A_1	1.20	−0.33	0.90	0.96
	A_2	−0.54	0.15	−0.39	−0.45
	A_3	−0.66	0.18	−0.51	−0.48
	a_{iso}				1.71
31 A	A_1	3.15	−0.63	2.55	2.61
	A_2	−1.44	0.30	−1.17	−0.72
	A_3	−1.71	0.33	−1.38	−1.89
	a_{iso}				0.69
20 B	A_1	6.42	−1.98	4.41	4.56
	A_2	−3.12	0.99	−2.13	−2.10
	A_3	−3.30	1.02	−2.28	−2.46
	a_{iso}				0.18
21 B	A_1	3.06	−1.02	2.04	2.28
	A_2	−1.44	0.51	−0.93	−0.96
	A_3	−1.62	0.51	−1.08	−1.32
	a_{iso}				0.03

[a] For numbering, see Fig. 34
[b] A_1, A_2, A_3 denote the traceless part of **A**
[c] For directions cosines, see[57]

6.1.2 Cu(II)-Doped Amino Acid Single Crystals

Cu(II) impurity complexes in amino acid single crystals have been the subject of several EPR studies[181–183]. Since nitrogen and proton hf structures are only partially resolved in the EPR spectra, no detailed information about the electronic properties of the complex in the neighborhood of the metal ion can be evaluated. ENDOR spectroscopy has therefore been applied[58, 63] to draw detailed pictures of the positions and the molecular environment of Cu(II) impurities in amino acid crystals.

Cu(II)-Doped ℓ-Alanine Single Crystals

The EPR spectra of Cu(II)-doped ℓ-alanine single crystals show a well resolved triplet structure which is due to a nitrogen ligand nucleus[182]. Moreover, in the partially

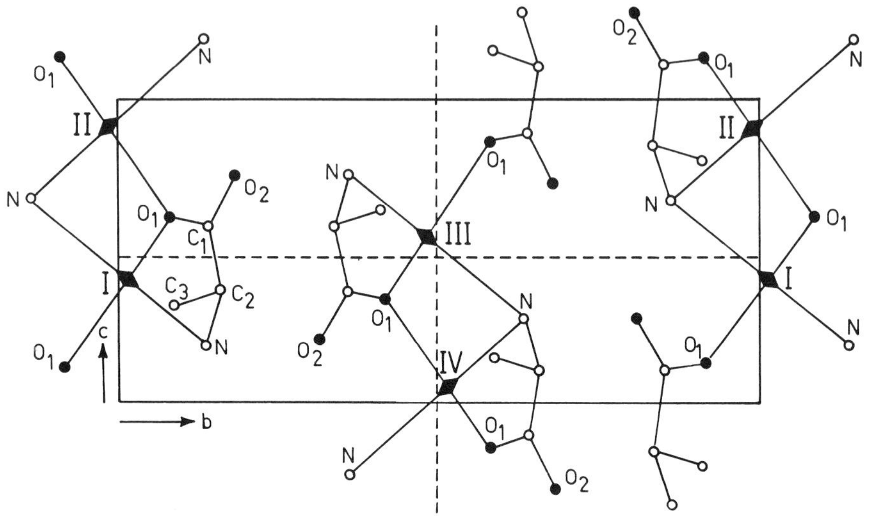

Fig. 35. Projection in the *bc* crystal plane of an ℓ-alanine single crystal. Proposed positions for the Cu(II) impurities are marked with *Roman letters*. (Adapted from Ref. 63)

deuterated crystal ℓ-alanine-d_4, the hf interaction of two amino protons is clearly resolved at 77 K, indicating that the stoichiometry of the amino group is $-NH_2$ rather than $-NH_3^+$ [183].

Figure 35 shows a projection of the crystal structure of ℓ-alanine on the *bc* plane. The four crystallographically equivalent positions of the Cu(II) impurity centers (indicated in the figure by Roman numerals) have first been proposed by Takeda et al.[184]. In this model one Cu(II) is bonded to *three* alanine molecules, with two oxygen and two nitrogen ligands in the first coordination sphere forming an almost planar but strongly distorted square.

The hfs and quadrupole tensors of one of the nitrogen ligands have been determined with ENDOR by Calvo et al.[63]. The ^{14}N-ENDOR transition frequencies observed between 11 and 23 MHz were found to depend significantly on the nuclear quantum number m_{Cu} of the EPR observer line. These shifts are due to Cu–N crossterms (Sect. 3.2) and amount to more than 1 MHz for certain orientations of $\mathbf{B}_0$. ENDOR resonances of

Table 9. Magnetic parameters of Cu(II) in ℓ-alanine (data from Calvo et al.[63]; A_i and Q_i in MHz)

Tensor	Principal values[a]	Tensor	Principal values[a]
g	$g_x = 2.049$ $g_y = 2.053$ $g_z = 2.257$	$\mathbf{A}^{N}$	$A_1 = 27.19$ $A_2 = 27.56$ $A_3 = 41.74$
$\mathbf{A}^{Cu}$	$A_1 = 51.6$ $A_2 = 109.2$ $A_3 = 594.6$	$\mathbf{Q}^{N\,b}$	$Q_1 = 0.097$ $Q_2 = 0.987$ $Q_3 = -1.084$

[a] Round figures, for errors and direction cosines, see[63]
[b] $e^2qQ/h = -2.170$ MHz, $\eta = 0.82$

the second nitrogen nucleus in the range between 6 and 10 MHz were observed but no magnetic parameters have been reported.

The EPR and nitrogen ENDOR data are summarized in Table 9. The principal directions of the **g** and $\mathbf{A}^{Cu}$ tensors are found not to be coaxial. This is particularly true for the smaller principal values. Comparison between EPR and X-ray data[185] shows that the normal on the plane spanned by the ligand nuclei differs by only ten degrees from the g_z-principal direction. Moreover, the axes of the largest coupling of $\mathbf{A}^N$ and $\mathbf{Q}^N$ lie approximately parallel to the N-N-direction. Thus, EPR *and* ENDOR data support Takeda's model[184] for the localization of Cu(II) in ℓ-alanine. The non-coaxiality of the tensors **g**, $\mathbf{A}^{Cu}$, $\mathbf{A}^N$ and $\mathbf{Q}^N$ indicates a low symmetry at the copper site with a complex ground state orbital for the unpaired electron.

The ^{14}N nuclear quadrupole data have been analyzed by Murgich et al.[186] in terms of the TD theory. The amino nitrogen was assumed to be tetrahedrally bonded to Cu(II), to one carbon and to two hydrogen atoms. In pure ℓ-alanine the principal axis of the EFG tensor lies approximately along the N–C bond, whereas in Cu(II)-ℓ-alanine this axis is oriented along a direction 14° away from the N–Cu bond. The corresponding quadrupole coupling constants vary from $e^2qQ/h = 1.184$ MHz ($\eta = 0.256$) to $e^2qQ/h = 2.170$ MHz ($\eta = 0.82$), respectively. This drastic change of the $\mathbf{Q}^N$ tensor upon metal complexation is explained by the higher electron affinity of Cu(II) with respect to C and H.

Cu(II)-Doped α-Glycine Single Crystals

The EPR spectra of Cu(II)-doped α-glycine single crystals have been studied by Windsch and Welter[181] and by Fujimoto and Janecka[182]. For the structure of the impurity two conceivable, crystallographically nonequivalent, Cu(II) sites A and B, sketched in Fig. 36, have been proposed[58]. In this model Cu(II) is coordinated to two glycine molecules and lies in the plane defined by two amino nitrogens and two carboxyl oxygens.

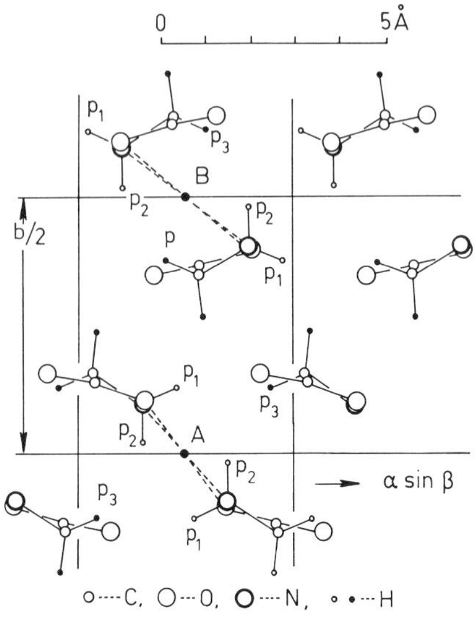

Fig. 36. Structure of α-glycine crystals. Projection along the *c*-axis. A and B are proposed sites for the Cu(II) impurities. (Adapted from Ref. 58)

Table 10. Magnetic parameters of $Cu(gly)_2$ in α-glycine (data from Fujimoto et al.[58]; A_i and Q_i in MHz)

Tensor	Principal values[a, b]	Tensor[c]	Principal values[a]
g	$g_x = 2.0434$ $g_y = 2.0715$ $g_z = 2.2644$	$\mathbf{A}^{H1}$	$A_1 = 6.35$ $A_2 = 1.99$ $A_3 = 1.14$ $a_{iso} = 3.16$
$\mathbf{A}^{Cu}$	$A_x = 156.4$ $A_y = 39.7$ $A_z = 468.7$	$\mathbf{A}^{H2}$	$A_1 = -11.10$ $A_2 = -5.63$ $A_3 = 6.13$ $a_{iso} = -3.50$
$\mathbf{A}^{N}$	$A_1 = 32.80$ $A_2 = 20.66$ $A_3 = 17.40$	$\mathbf{A}^{H3}$	$A_1 = -9.75$ $A_2 = -13.92$ $A_3 = 4.73$ $a_{iso} = -6.32$
$\mathbf{Q}^{N}$ [d]	$Q_1 = 1.26$ $Q_2 = -1.81$ $Q_3 = 0.55$		

[a] For direction cosines, see[58)]
[b] The tensor **g** and $\mathbf{A}^{Cu}$, and $\mathbf{A}^{N}$ and $\mathbf{Q}^{N}$ are found to be nearly coaxial
[c] H1: methylene proton; H2, H3: amino protons
[d] After extraction of a trace of $\Sigma Q_i/3 = 0.3$ MHz (sic!)

Proton and nitrogen ENDOR data on $Cu(gly)_2$ have been reported by Fujimoto et al.[58)]. The second order splittings in the ^{14}N-ENDOR spectra are due to the inversion-symmetry of the impurity complex and have been analyzed by a treatment similar to that discussed in Sect. 3.2. The double-line structure of the proton ENDOR lines observed for some $\mathbf{B}_0$ field orientations was assumed to originate from an *indirect* coupling between equivalent protons. However, a numerical calculation shows that such an interaction is by far too small to be responsible for the observed splittings of up to 50 kHz[40)]. Instead, the doublets must be described by a proton dipole-dipole coupling which is not related to symmetry arguments (Sect. 3.2).

EPR and ENDOR data are summarized in Table 10. In partially deuterated glycine crystals the intensities of proton H2 and H3 are reduced, implying that $Cu(gly)_2$ involves two pairs of exchangeable amino protons. The assumption that H2 and H3 are geminate protons has been verified by a nuclear spin decoupling experiment[40)]. All the hfs tensors of the measured protons are strongly rhombically distorted, indicating that the unpaired electron is quite delocalized in this system. The Cu–H distances estimated from the anisotropic parts of the hfs tensors using the point-dipole approximation were found to be in reasonable agreement with the distances calculated from X-ray data for the proposed site A. Thus, ENDOR allows discrimination between the Cu(II) sites A and B depicted in Fig. 36.

6.1.3 Cu(II)-Doped Organic Ferroelectric Crystals

Cu(II)-Doped Triglycine Sulfate

Several ENDOR investigations on X-irradiated, Cu(II)-, and VO(II)-doped single crystals of triglycine sulfate (TGS) in its ferroelectric phase have been reported by Windsch and coworkers[60, 187–189]. The unit cell of pure TGS contains three types of glycine molecules (glycine I and III: $NH_3^+CH_2COOH$ (glycinium ions), and glycine II: $NH_3^+CH_2COO^-$ (*zwitterion*)[190]). Upon doping with Cu(II), $Cu(gly)_2$ complexes with glycines II and III (Fig. 37) are formed.

Figure 38 shows the angular dependence of the ENDOR transitions of the amino protons in Cu(II)-doped TGS for $\mathbf{B}_0$ normal to a'[188]. The observed eight-line pattern

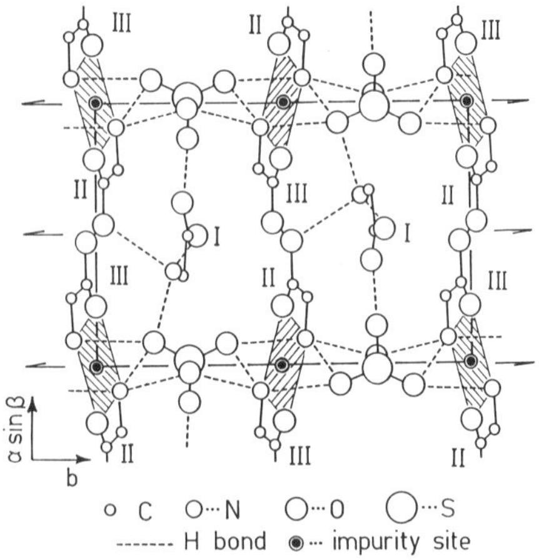

Fig. 37. Structure of TGS. Projection along the *c*-axis. (Adapted from Ref. 190)

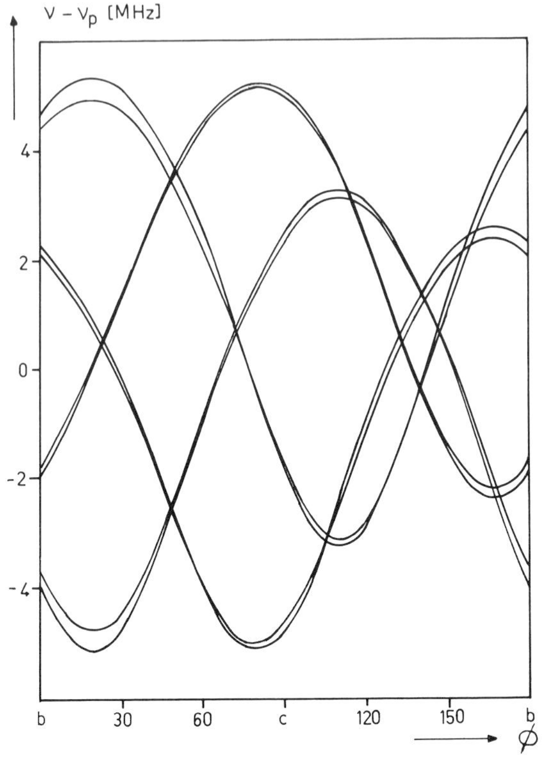

Fig. 38. Angular dependence of the amino proton ENDOR signals of TGS doped with Cu(II) ions. Rotation axis along a'. (Adapted from Ref. 188)

clearly indicates that the couplings of the amino protons in the two Cu(II)-chelates are different from each other. Different hfs tensors have also been found for the methylene groups of corresponding protons in glycine II and III. Cu–H distances of the methylene protons obtained from X-ray data are in good agreement with those calculated from the dipolar part of the hfs tensors (point-dipole approximation). Owing to the short distances between the NH_2 protons and the Cu(II) ions it was not possible to determine the positions of these protons from the anisotropic part of the hfs tensors.

The observed splittings of the nitrogen ENDOR lines, which vary from −230 to 430 kHz for a rotation of the crystal around *a'*, have also to be interpreted as originating from two *nonequivalent* nitrogen nuclei rather than from a second order splitting related

Table 11. Proton and nitrogen hfs data of $Cu(gly)_2$ and $VO(gly)_2$ in TGS (data from Böttcher et al.[60, 188, 189]; in MHz)

Nucleus		Cu(II) A_1, A_2, A_3	a_{iso}	VO(II) A_1, A_2, A_3	a_{iso}	distance[a]
amino[b]		−6.30		14.97		
protons		6.15	3.75	1.33	5.47[c]	2.55
		11.30		0.67		
		−6.60		8.52		
		6.50	4.05	−3.37	0.41	2.67
		12.15		−3.95		
		8.80		10.77		
		13.50	5.75	−0.23	3.01[c]	2.65
		−5.00		−1.52		
		8.95		8.10		
		13.80	5.70	−3.99	−0.21	2.71
		−5.50		−4.75		
methylene[b]	H 8[d]	4.35		5.80		
protons		−1.80	0.07	−1.28	0.48	3.07
		−2.30		−3.07		
	H 9	6.93		5.05		
		2.08	3.65	−2.70	0.15	3.18
		1.79		−1.90		
	H 13	4.44		2.15		
		−1.86	0.08	−1.00	0.07	4.24
		−2.36		−0.95		
	H14	7.27		2.05		
		2.10	3.75	−0.95	0.07	4.30
		1.73		−0.90		
N[e]		6.69	16.8[f]	–[g]		
		−3.45	−8.4			
		−3.24	−8.4			

[a] Central ion-proton distance in Å
[b] For direction cosines, see[60, 188)]
[c] Geminal protons
[d] For notation, see[188)]
[e] Dipolar part $A_i^N - a_{iso}^N$ [189)]
[f] From an EHT-SCCC calculation[189)]
[g] No hf data available

to inversion-symmetry[189, 191]. This is supported by the fact that this doublet splitting is *not* dependent on the EPR observer line[191], as is found for two magnetically equivalent I = 1 nuclei[62].

The hf parameters of the protons in the NH_2 and CH_2 groups and the mean values of the nitrogen hf coupling constants are listed in Table 11. It should be noted that the hfs data of the amino protons in TGS strongly deviate from the values found for Cu(II)-doped α-glycine crystals[58] (Table 10). This indicates that the structures of the $Cu(gly)_2$ complexes in α-glycine and in TGS are significantly different from each other.

Proton and nitrogen ENDOR data confirm that the symmetry of $Cu(gly)_2$ in TGS in its ferroelectric phase is only C_1. The differences in the observed hfs tensors of corresponding nuclei in glycine II and III, however, are not very pronounced, so that deviations from the C_i or C_{2h} symmetry should be small.

The magnetic parameters have been interpreted using EHT-SCCC calculations[189, 192]. If C_{2h} symmetry is assumed, the **g** tensor and the anisotropic part of the hfs tensors $\mathbf{A}^{Cu}$ and $\mathbf{A}^{N}$ could satisfactorily be explained. According to these calculations the largest principal axis of the nitrogen hfs tensor lies in the complex plane and deviates 27° from the Cu-N direction (compared with 17° from ENDOR data)[6].

6.1.4 *Copper-Containing Proteins*

Since the first EPR work on Cu(II) ions in proteins in the late fifties[193], a great many EPR investigations on copper-containing proteins have been reported[194–198]. For a classification of the copper proteins into type I (blue copper), type II (non-blue copper) and type III (binuclear cupric pair), the reader is referred to Fee[197].

In frozen solution EPR spectra of copper proteins, ligand and copper hf interactions are only partially resolved, or not at all. It is therefore striking that only few ENDOR studies on copper proteins have been published so far[17, 199–201]. Nevertheless, these few ENDOR results already demonstrate the power of the double resonance technique to probe the coordination environment of the copper-containing sites in these types of proteins.

Stellacyanin

The blue copper protein stellacyanin, with a molecular weight of about 20,000, is obtained from the Japanese lacquer tree *Rhus vernicifera*. The EPR spectrum is described by roughly axial **g** and $\mathbf{A}^{Cu}$ hfs tensors and an unusually small a_{iso}^{Cu} value. As shown in Fig. 39 a, only the largest copper hf value A_x^{Cu} can be directly determined from the EPR spectrum[202]. This coupling does *not* lie along the largest g-principal axis, in contrast to the usual behaviour of square planar copper complexes.

---

6 A similar ENDOR study on VO(II)-doped TGS[60] will be discussed in Sect. 6.6

Two ENDOR studies on stellacyanin in frozen solution have been published[17, 199]. In the initial work by Rist et al.[17], ENDOR transitions in the 10–26 MHz region have been observed. To discriminate proton against nitrogen resonances, the spectra were measured at the two microwave frequencies 8.58 GHz and 9.56 GHz. This corresponds to a magnetic field change of 10% and thus to a shift of the proton and nitrogen lines of about 1.3 MHz and <0.1 MHz, respectively. From these shifts it has been established that at least one ligand of stellacyanin is a nitrogen. This conclusion has recently been confirmed by Roberts et al.[199] who reported quantitative results obtained from nitrogen and copper ENDOR spectra.

Besides a strongly coupled proton with a nearly isotropic hfs of $|A^H| \simeq 20$ MHz, two clearly separated nitrogen peaks between 15–26 MHz with unresolved quadrupole and nuclear Zeeman splittings have been observed along all three turning points of the **g** tensor (Fig. 39b). In the evaluation of the hf data collected in Table 12.1 it is assumed that the two metal-coordinated ^{14}N exhibit roughly axial hfs tensors with the $A_{\|}^{N}$ values oriented approximately perpendicular to each other.

From the ESE envelope modulation pattern of stellacyanin it was argued that at least one histidine is bound to copper[203]. The close similarity of the ENDOR data of the two nitrogen ligands suggests that both nitrogens may belong to coordinated histidines. This assignment is supported by X-ray diffraction data on the blue copper proteins plastocyanin and azurin[204, 205], where the copper is coordinated by nitrogens from two histidines and by two sulfur atoms.

As mentioned above it is not possible to identify the number of magnetically equivalent nitrogens from first order expressions of the ENDOR transition frequencies. Since

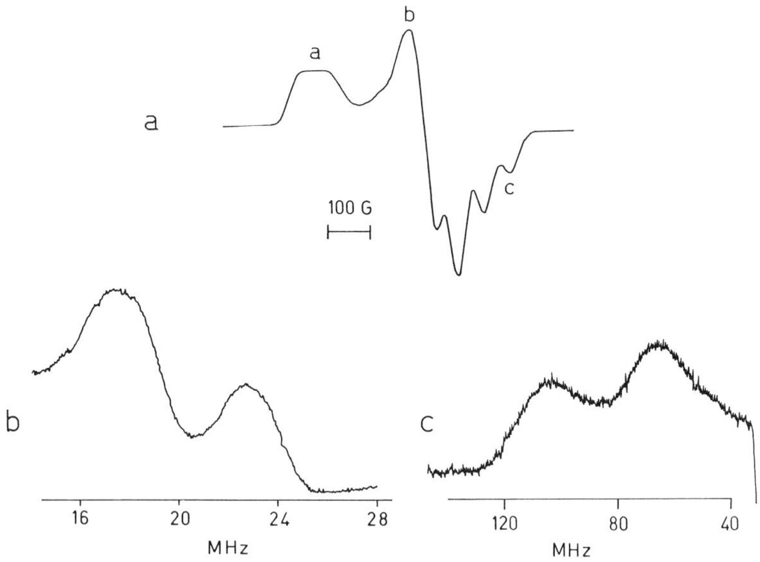

Fig. 39 a–c. EPR and ENDOR spectra of stellacyanin. **a)** Experimental EPR spectrum. The fields near which the single crystal-like ENDOR spectra were measured are marked with (**a**)g_z, (**b**)g_y and (**c**)g_x; temperature 77 K. **b)** ^{14}N-ENDOR spectrum with $\mathbf{B}_0$ along g_z; temperature 2 K. **c)** Cu-ENDOR spectrum with $\mathbf{B}_0$ near position (**c**) (opposite phase to ^1H- and ^{14}N-ENDOR). (Adapted from Ref. 199)

Table 12.1. Magnetic parameters of the blue copper protein stellacyanin (data from Roberts et al.[199]; A_i and Q_i in MHz)

	Axes x	y	z
g	2.018	2.075	2.282
$\|A^{Cu}\|$	167	87	96
$\|Q^{Cu}\|$	6	4	10
$\|A^{N}\|$	$A_\perp(1) \simeq 32$ $A_\parallel(2) \simeq 44$	$A_\perp(1) = A_\perp(2)$ $\simeq 32$	$A_\parallel(1)^a \simeq 44$ $A_\perp(2) \simeq 32$

[a] The numbers in parenthesis correspond to the two nitrogen ligands

Table 12.2. EPR and Cu-ENDOR parameters of cytochrome *c* oxidase; A_i^{Cu} in MHz

	Axes x	y	z
g^a	1.99	2.03	2.185
$\|A^{Cu}\|^b$	68	98	90
$\|A^{Cu}\|^a$	60	75	90
$\|A^{Cu}\|^c$	~ 28	127	≤ 122

[a] X-band EPR data from[212]
[b] Cu-ENDOR data from[200]
[c] S-band EPR data from[210]

the intense blue color of this class of copper proteins is probably due to a sulfur-copper charge-transfer transition[206], it might be possible that the copper is ligated by three nitrogens and one sulfur.

Weak Cu-ENDOR signals detected with rapid frequency sweep rates ($d\nu/dt \geq 10$ MHz/s) have been observed at $\mathbf{B}_0$ fields corresponding to the three principal axes directions of the **g** tensor and to numerous intermediate values[199]. The Cu-ENDOR spectrum obtained with $\mathbf{B}_0$ along g_x (position c) is shown in Fig. 39c. The transitions are about 20 MHz broad and are observed as enhanced or decreased EPR absorptions, depending on experimental conditions. The best estimates for the hf and quadrupole parameters obtained from combined results of ENDOR and EPR simulations are collected in Table 12.1. Roberts et al.[199] explain the unusual copper hf data by assuming a flattened tetrahedral geometry of the chromophore. Since in this geometry the odd electron orbital is a mixture of $3d_{xy}$ and $4p_z$, a_{iso}^{Cu} is expected to be reduced and the minimum hf component to point along the maximum g-value. The authors believe that the p-orbital mixing related to this geometry is entirely sufficient for an explanation of the copper hfs tensor and that the nature of the ligand is of minor importance. In particular, the sulfur ligand, which produces the blue color, is not required for an understanding of the observed magnetic properties.

The highly rhombic nuclear quadrupole tensor of the Cu(II) ion indicates that the *total* electron density at the copper site deviates considerably from any idealized symmetry.

Cytochrome *c* Oxidase

The protein cytochrome *c* oxidase, with a monomeric molecular weight of about 100,000, contains two a-type hemes and two copper atoms[207]. A low-spin ferric heme signal and a so-called "intrinsic copper" signal near g = 2.0 are observed in the frozen solution EPR spectrum of the fully oxidized oxidase. The latter signal shows EPR features which one could assign either to a thiyl radical (R-S·)[208] or to a cupric ion center with an unusually low $g_{\|}$ value[209]. Since no hf structure from copper is resolved at X- or Q-band frequencies, the controversy could not be solved unambiguously from EPR data alone.

Recently, hf structure associated with the copper signal of cytochrome *c* oxidase has been reported by Froncisz et al.[210] which used *octave bandwidth S-band EPR spectroscopy* (2–4 GHz). The observed structure has been attributed to copper hfs and to an additional magnetic interaction. Data obtained from powder simulation of the EPR spectra at 2.62 GHz and 3.78 GHz are collected in Table 12.2. In a subsequent paper Froncisz and Hyde[211] have shown that in S-band EPR spectra of copper complexes in frozen solutions, *improved spectral resolution* can be achieved. This new technique, which allows a proper selection of the microwave frequency between 2 and 4 GHz, is therefore recommended for studying powder EPR spectra of these types of compounds.

Proton and nitrogen ENDOR from the "intrinsic copper" signal in cytochrome *c* oxidase was first reported by van Camp et al.[201]. As in the ENDOR work on stellacyanin[17] the transitions could be assigned to the corresponding types of nuclei by using two different microwave frequencies (9.00 and 9.39 GHz). Only *one* pair of nitrogen lines split by $2\nu_N$ have been observed along all principal directions of the **g** tensor. Thus, the nitrogen quadrupole coupling is either nearly zero (two-line ENDOR pattern), or the transitions of the second pair of lines are obscured near the free proton frequency. In the former case, the nitrogen frequencies correspond to an isotropic hf coupling constant of about 17 MHz. For the two strongly coupled protons, isotropic hf values $|a^H_{iso}|$ of 12 and 19 MHz have been found. It is supposed that these protons, which are not exchangeable in D_2O, are attached to carbons that are one bond away from the point of ligation to copper[201].

In 1980 Hoffman et al.[200] published the first Cu-ENDOR data on cytochrome *c* oxidase, which unambiguously established that the "intrinsic copper" signal in this protein is indeed due to copper, which is part of the oxidation-reduction center. As in stellacyanin[199] the rapid scanning technique had to be applied to detect ENDOR from the copper ion. Along all three principal axes of the **g** tensor only a single broad peak was observed, so that the copper quadrupole data could not be obtained from the ENDOR spectra. No evidence for a hf coupling greater than 19 MHz in the ENDOR spectrum was found, in contradiction to the findings of Froncisz et al.[210], which had to use an additional hf interaction of about 70 MHz in the simulation of the S-band EPR spectra. The data obtained from Cu-ENDOR are comparable to previously reported computer simulations of X-band EPR spectra[212] (Table 12.2).

The unusual **g** and $\mathbf{A}^{Cu}$ tensors may both be reproduced by appropriate mixing of 4p- and 3s-orbitals into the 3d manifold. A flattened-tetrahedral, rhombically distorted stereochemistry about the Cu(II) ion is in agreement with the magnetic parameters given in Table 12.2[212]. Hoffman et al.[200] argue that a Cu(II) ion ligated by nitrogen(s) and two or more mercaptide sulfurs would explain both the observed magnetic properties and the suggested high degree of covalency[213].

The authors also examined a model which describes the paramagnetic site in terms of a Cu(I) ion coordinated to a sulfur π radical[213, 214]. In the discussion of the magnetic data, however, they show that the proton hf parameters and the **g** tensor found for cytochrome *c* oxidase are not consistent with a thiyl radical ligated to a Cu(I) center.

6.1.5 *Other Cu(II) Complexes*

Cu(II)-Doped Zinc Acetate Dihydrate

EPR and proton ENDOR measurements on Cu(II)-doped zinc acetate have been reported by Atherton and Horsewill[215]. In contrast to the pure copper salt, which consists of copper pairs, zinc acetate has a monomeric structure in which two water molecules are coordinated to each zinc ion[216], as illustrated in Fig. 40. The magnetic parameters are summarized in Table 13. Both the **g** and $\mathbf{A}^{Cu}$ tensor obtained from single crystal and

Table 13. Magnetic parameters of Cu(II)-doped zinc acetate dihydrate (data from Atherton and Horsewill[215]; A_i and a_{iso} in MHz)

EPR data

	x	y	z
g	2.091	2.095	2.473
$\mathbf{A}^{Cu}$	69	69	260

ENDOR data

Proton[a]	a_{iso}	traceless components[b] of $\mathbf{A}_d$
1	2.61	9.88 −6.37 −3.50
1′	2.61	9.89 −6.40 −3.49
2	−1.26	9.87 −6.48 −3.40
2′	−1.30	9.82 −6.48 −3.34
3	−0.23	4.79 −2.75 −2.04
3′	−0.25	4.77 −2.71 −2.05

[a] From numbering, see Fig. 40

[b] For direction cosines with respect to the crystal axes, see[215]

powder samples have axial symmetry within experimental accuracy and are assumed to be coaxial. The principal axis of $g_{\parallel}$ forms an angle of 15° with the Cu-O_I direction, if the Cu(II) ions substitute the Zn(II) ions without changing the geometry of the ligand molecules.

Six proton tensors have been analyzed independently. The direction cosines indicate that the protons are *geometrically equivalent* in pairs, as expected from the crystal struc-

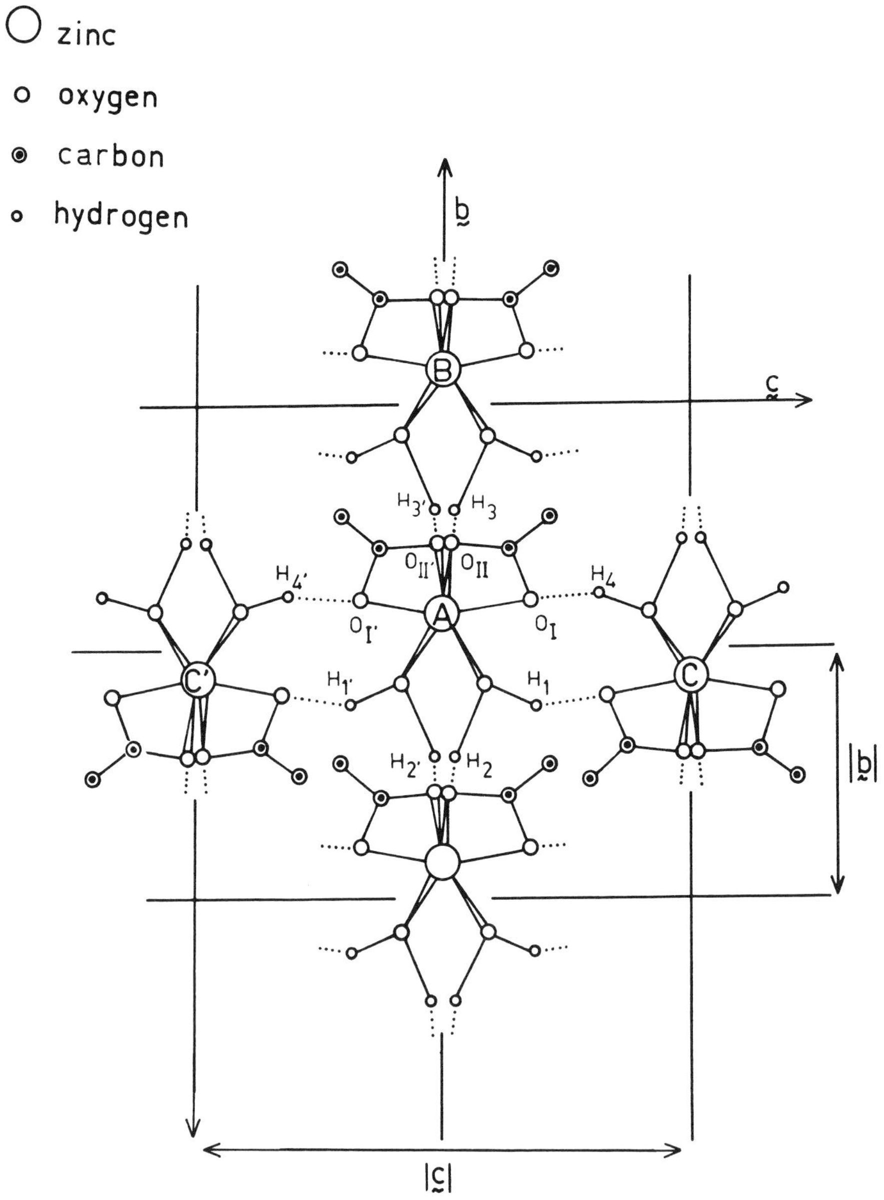

Fig. 40. Projection of the zinc acetate structure on the *bc* plane. (From Ref. 215)

ture. Using the simple point-dipole formula (5.6), protons 1 and 2 as well as 1′ and 2′ are found to belong to the water molecules coordinated to the cupric ion, whereas protons 3 and 3′ have been identified as the protons 2 and 2′ of the adjacent zinc complex shifted by one unit cell lengths along *b*. For water protons in nearest-neighbor molecules along the *c* direction (H 4 and H 4′ in Fig. 40), approximate hfs parameters have been obtained by combining single crystal and powder ENDOR data. The hf coupling of H 4 (H 4′) may be roughly characterized by a traceless axial tensor with $A_{\parallel}^{H} \simeq 3$ MHz. The orbital contributions to the proton interactions (Sect. 5.1) have been evaluated to be $|A_{pseudo}| < 0.3$ MHz.

It is supposed that the skewed orientation of the water molecules in copper acetate dihydrate is responsible for the different spin densities on the hydrogen atoms 1 and 2. The balancing of the resulting spin densities against those arising from spin polarization leads to net results of opposite sign for the two protons.

Cu(II)-bis(2,4-pentanedionate), $Cu(acac)_2$

Single crystal Cu-ENDOR spectra of $Cu(acac)_2$ have been studied by Kita et al.[158] up to 100 MHz. The ENDOR transitions of the isotopes ^{63}Cu and ^{65}Cu are found to be well separated for most orientations of $\mathbf{B}_0$. The angular dependence of some Cu-ENDOR frequencies shows pronounced double-minima instead of a single minimum in the region where the hf coupling is smallest. Similar rotation patterns have been observed in Co-ENDOR spectra of Co(acacen) (Fig. 43 b).

The nuclear quadrupole tensor of the Cu(II) ion is discussed in terms of the formalism outlined in Sect. 5.2.1. Kita et al.[158] treated the Sternheimer antishielding factor $(1 - \gamma_\infty)$ as an adjustable parameter. The rhombicity of the measured quadrupole tensor was found to originate from the charge distribution on the ligands, the field gradient due to the copper valence electrons being nearly axially symmetric.

As illustrated in Fig. 41 (see p. 84), the number of observed Cu-ENDOR lines (^{63}Cu enriched sample[217]) does not follow the usual ENDOR selection rules for EPR observers with resolved hfs (two transitions for EPR observers with $m_I = \pm I$, four transitions for EPR observers with $-I < m_I < I$; Sect. 3.2). This may be due to transfer of microwave saturation to EPR transitions other than the one which is used as an observer. A corresponding effect has been observed in Co-ENDOR spectra of Co(acacen)[61].

6.2 Co(II) Compounds (d^7)

In the last few years, EPR has been widely used to study the electronic structure of four- and five-coordinated low-spin Co(II) complexes. Compounds of this class gained considerable interest because of their relation to biological oxygen carriers and to vitamin B_{12r}. The ground state of the five-coordinated complexes is accepted to be $|d_{z^2}\rangle$, in contrast to the four-coordinated compounds for which the determination of the correct ground state was quite troublesome, due to the fact that these types of Co(II) complexes have low-lying excited states.

For a comprehensive summary of the electronic structure of Co(II) complexes with *Schiff* bases and related ligands, the reader's attention is directed to a recent review by Daul et al.[218].

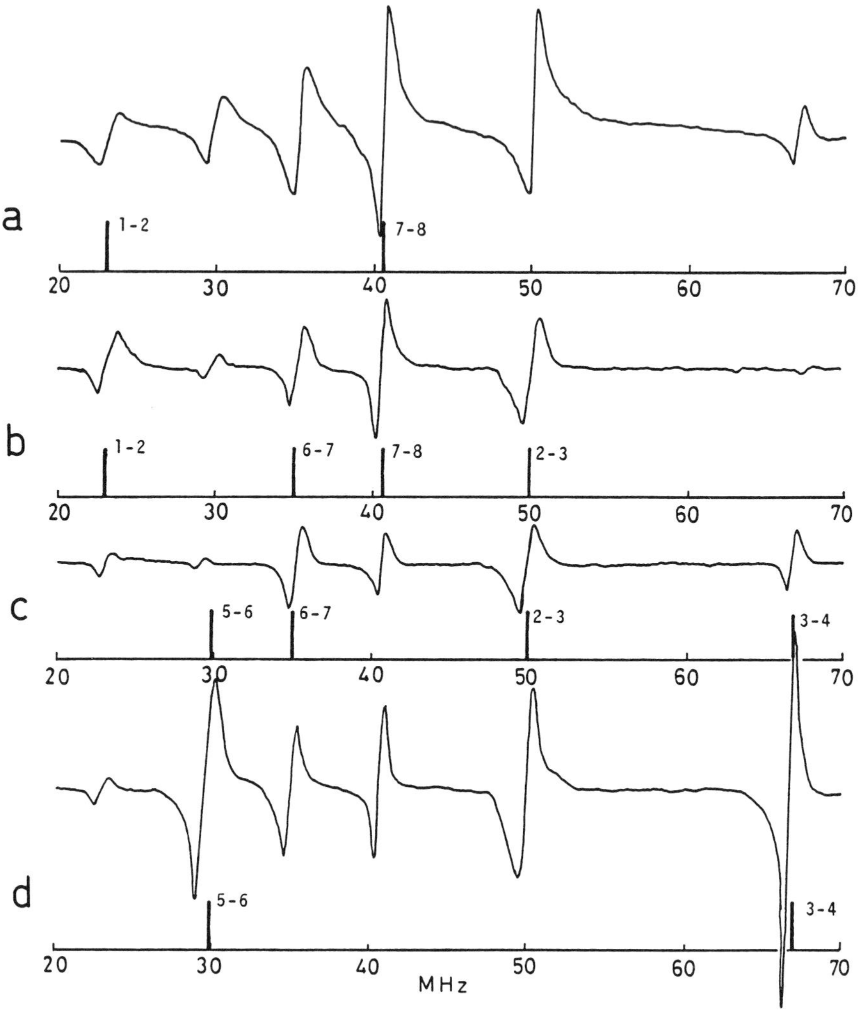

Fig. 41 a–d. Cu-ENDOR spectra of Cu(acac)$_2$. EPR observers: **a)** 1–8, **b)** 2–7, **c)** 3–6, **d)** 4–5 EPR transition. (From Ref. 217)

[N,N′-ethylene-bis(acetylacetonatiminato)]Co(II), Co(acacen)

No information concerning the distribution of the unpaired electron on the chelate ligands of four-coordinated low-spin Co(II) *Schiff* base complexes is available from single crystal EPR spectra. In particular, no nitrogen hf interaction, which is sensitive to the ground state configurations, is observed.

A single crystal ENDOR study comprising ^{1}H-, ^{13}C-, ^{14}N- and ^{59}Co-ENDOR investigations of the *Schiff*base compound Co(acacen) (Fig. 42) doped into Ni(acacen) · 1/2 H_2O was therefore undertaken by Rudin et al.[12, 59, 61, 219]. As illustrated in Fig. 2b, ENDOR spectroscopy allows hf and quadrupole interactions of all the magnetic ligand

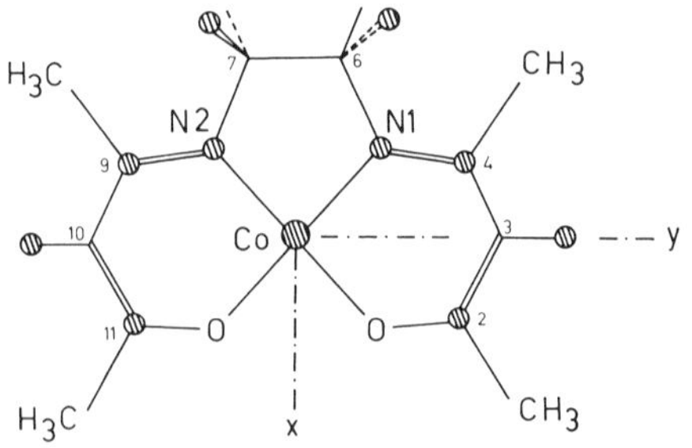

Fig. 42. Structure of Co(acacen). Circles mark nuclei for which the magnetic parameters have been analyzed. (Ref. 12)

Table 14. Magnetic parameters of Co(acacen) in Ni(acacen) · $1/2\,H_2O$ (data from Rudin et al.[12, 59, 61]; in MHz)[a, b]

Atom	A_1	A_2	A_3	a_{iso}	Q_1	Q_2	Q_3
Co	389.4	-93.9	96.30	130.6	-2.76	2.61	0.15[c]
N 1	-4.75	-3.75	-2.10	-3.53	0.85	-0.73	-0.12[d]
N 2	-4.95	-3.96	-1.93	-3.61	0.89	-0.72	-0.20[e]
C 1	-6.39	-2.96	-0.28	-3.21			
C 2	-6.46	-3.16	-0.21	-3.28			
C 3	-4.56	-3.02	-0.53	-2.70			
C 4	-4.36	-3.08	-0.57	-2.67			
H 3[f]	2.93	-1.10	-1.83	-1.31			
H 6 ax[f]	4.03	-2.00	-2.03	3.90			
H 7 ax[f]	4.17	-1.84	-2.33	4.40			
H 10[f]	3.00	-1.11	-1.89	-1.34			

[a] For direction cosines, see[12, 59, 61]
[b] $g_x = 3.227$, $g_y = 1.908$, $g_z = 1.987$[61]
[c] $e^2qQ/h = -116.46$ MHz, $\eta = 0.89$
[d] $e^2qQ/h = 1.74$ MHz, $\eta = 0.72$
[e] $e^2qQ/h = 1.78$ MHz, $\eta = 0.58$
[f] principal values of T, Eq. 5.14

nuclei in Co(acacen) to be resolved. The analyzed EPR and ENDOR parameters are collected in Table 14.

The first order EPR spectrum of Co(acacen) consists of two sets of eight "allowed" $\Delta m_S = \pm 1$, $\Delta m_{Co} = 0$ transitions ($I = 7/2$, two magnetically nonequivalent sites). This simple pattern, however, was only observed for orientations of $\mathbf{B}_0$ near the principal axis g_x. If $\mathbf{B}_0$ lies near the plane spanned by g_y and g_z, "forbidden" $\Delta m_S = \pm 1$, $\Delta m_{Co} = \pm 1$, 2 transitions occur (Fig. 2a).

Ligand ENDOR[12, 59, 219]. The magnetic tensors of all nuclei marked by a circle in Fig. 42 have been determined. The tensors are approximately related in pairs by a two-fold symmetry axis (molecular x-axis), (e.g. $\mathbf{A}^{H3} = \tilde{\mathbf{\Gamma}}(C_2(x))\mathbf{A}^{H10}\mathbf{\Gamma}(C_2(x))$. Thus, Co(acacen) in Ni(acacen) · $1/2H_2O$ has nearly $C_2(x)$ symmetry, in agreement with X-ray data[220]. The largest hf couplings of the methylene protons H 3 and H 10 as well as of the axial protons H 6ax and H 7ax in the ethylenediamine bridge are oriented along the corresponding Co-H directions. Due to the small Fermi contact terms, the ENDOR frequencies of the two equatorial protons H 6eq and H 7eq are hidden for most crystal settings by the large number of proton transitions near the free proton frequency ν_p (methyl protons of Co(acacen) and protons of the host compound), so that their hfs tensors could not be analyzed from the rotation patterns. From the isotropic couplings of H 3 and H 10 the $2p_z$-spin densities on the adjacent carbons C 3 and C 10 have been calculated to about 2.2% by using the McConnell relation[221].

The ^{14}N hf and quadrupole parameters observed in Co(acacen) by Rudin et al.[59] are the first magnetic data reported on equatorial nitrogen ligand nuclei in a low-spin Co(II) complex. Only two of the four predicted $\Delta m_N = \pm 1$ ENDOR transitions (3.9) were observed for each nitrogen nucleus. A numerical calculation of the transition probabilities shows that the corresponding transitions in the other m_S-state are at least ten times less intense (hyperfine enhancement).

Besides the "allowed" $\Delta m_N = \pm 1$ transitions, "forbidden" $\Delta m_N = \pm 2$ lines were observed for certain magnetic field orientations (Sect. 3.3). The latter transitions proved to be useful for the assignment of the $\Delta m_N = \pm 1$ lines to the corresponding nitrogen nuclei. Since the assumption $\mathcal{H}_{hfs} \gg \mathcal{H}_Q$ (Sect. 3.2) is no longer well fulfilled for the small nitrogen hf couplings, the tensors $\mathbf{A}^N$ and $\mathbf{Q}^N$ have been evaluated by numerical diagonalization of the spin Hamiltonian matrices.

The two nitrogen hfs tensors in Co(acacen) are not exactly geometrically equivalent (Table 14). Similar deviations have also been observed in the corresponding copper complex Cu(acacen)[222] and in Cu(salen)[173]. For both nitrogen ligands, the principal axes A_2^N lie approximately along the Co–N bond. The negative isotropic coupling of $\simeq -3.5$ MHz implies an s-spin density $f_s^N = -0.0023$. This value is in good agreement with the spin density of -0.0018 calculated with the INDO method by Fantucci et al.[223]. The remaining dipolar part of $\mathbf{A}^N$ arises mainly from positive spin density in the $2p_z$-orbital, supporting the $|^2A_2, yz\rangle$ ground state proposed for four-coordinated *Schiff* base Co(II) complexes[218] (nonzero overlap of the orbitals $3d_{yz}$ and $2p_z$).

^{13}C hfs tensors have been measured for the carbons at positions 2, 4, 9 and 11 (90% ^{13}C-enriched). The tensors are strongly rhombically distorted with the largest coupling oriented approximately along the molecular z-axis (Table 14). Up to now the significantly different pairs C 1, C 2 and C 3, C 4 could not be assigned to carbons adjacent either to nitrogens (4, 9) or to oxygens (2, 11) on the basis of spectroscopic arguments.

A detailed interpretation of the magnetic parameters of the ligand nuclei of Co(acacen) using an extended Hückel-type calculation has been given by Rudin et al.[12, 219].

Co-ENDOR[61]. Co-ENDOR transitions in the frequency range 40 to 220 MHz have been detected using the high-field EPR line $|m_{Co}| = 5/2$ as observer. The ENDOR spectrum consists of the four transition frequencies $\nu(-7/2 \leftrightarrow -5/2;\ m_S = \pm 1/2)$ and $\nu(-5/2 \leftrightarrow -3/2;\ m_S = \pm 1/2)$. Linewidths between 200 kHz and 1 MHz have been found depending on crystal orientation, and on the magnetic quantum numbers m_{Co} and m_S. These values are considerably larger than those found for the ENDOR linewidths of the ligand nuclei ($\gtrsim 10$ kHz for ^{14}N, $\simeq 30$ kHz for ^1H).

Figure 43 a shows the angular dependence of the four Co-ENDOR transitions for rotations of the crystal around three cartesian axes. For rotation I, the angular dependence shows a noteworthy peculiarity near $\varphi = 0$ (Fig. 43 b). A similar behavior has already been mentioned for Cu-ENDOR spectra of $Cu(acac)_2$[158]. Because of the large and strongly anisotropic cobalt hf interaction, the ENDOR frequencies obtained with the

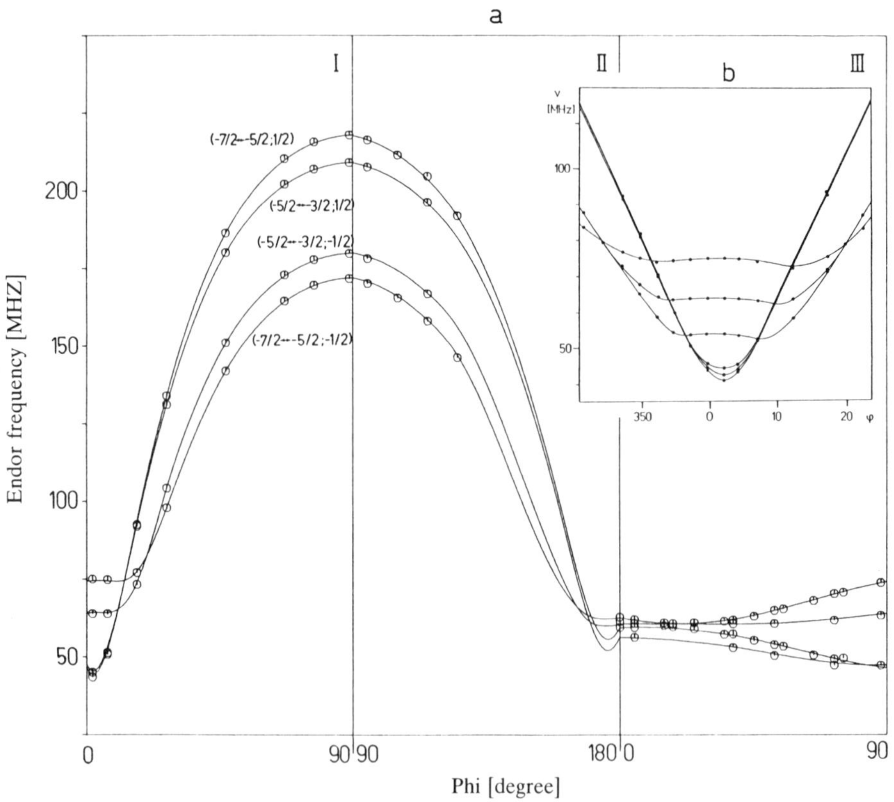

Fig. 43 a, b. Angular dependence of the Co-ENDOR transitions in Co(acacen); temperature 8 K. **a)** Rotation pattern for rotations around the axes I, II and III. The labels ($m_{Co} \leftrightarrow \overline{m}_{Co}$, m_S) denote the transitions between the energy levels E(m_S, m_{Co}) and E(m_S, $\overline{m}_{Co}$). Only half of the measured interval for each rotation is shown. (From Ref. 61). **b)** Rotation I near $\varphi = 0$. (From Ref. 219)

perturbation approach summarized in Sect. 3 differ significantly from the correct values. As for the nitrogen ligands, the hfs and quadrupole interactions of the cobalt nucleus (Table 14) have been determined from the ENDOR data by numerical diagonalization of the spin Hamiltonian matrices. The principal axes systems of **g**, $\mathbf{A}^{Co}$ and $\mathbf{Q}^{Co}$ do not coincide. The principal axes g_x, A_1^{Co} and Q_1^{Co}, however, are practically parallel and point along the C_2(x) axis of the Co(acacen) molecule. This again indicates that the molecular symmetry is C_2 rather than C_{2v}, although the deviations from C_{2v} are quite small. The orientations of A_2^{Co} and A_3^{Co}, (Q_2^{Co} and Q_3^{Co}) which include an angle of $-12 \pm 3°$ ($-10 \pm 5°$) with the corresponding **g** tensor axes might be influenced by the lattice, possibly by the water of crystallization.

Further information available from the Co-ENDOR spectra concerns the signs of the $\mathbf{A}^{Co}$ principal values obtained from higher order shifts. Since the sign of A_1^{Co} is unambiguously known from theoretical arguments[218], the absolute signs of all the principal values of $\mathbf{A}^{Co}$ and $\mathbf{Q}^{Co}$ are known. The *negative* sign found for A_2^{Co} clearly favors the $|^2A_2, yz\rangle$ ground state[224, 225]. The experimental **g** and $\mathbf{A}^{Co}$ tensors are in good agreement with the corresponding data obtained with different theoretical approaches[218, 224–226]. Using the third order perturbation theory developed by McGarvey[226] and Hitchman[225], the parameters $\Delta = E(^2A_1, z^2) - E(^2A_2, yz) = 2000\ cm^{-1}$, $\lambda = -400\ cm^{-1}$ (spin orbit coupling constant), $P = 2\mu_B g_{Co}\mu_N \langle r^{-3}\rangle_{3d} = 0.021\ cm^{-1}$ and $\varkappa = 0.24$ (isotropic hf coupling, in units of P) have been calculated from the **g** and $\mathbf{A}^{Co}$ principal values.

The cobalt quadrupole coupling constants have been discussed in terms of central ion and ligand orbital populations, as outlined in Sect. 5.2.1. The MO data for the computation of the EFG were obtained from a SCCC-MO calculation[227]. With the shielding factors $\gamma_\infty = -9$ and $R_{3d} = 0.2$ and an $\langle r^{-3}\rangle_{3d}$ value of 5.05 a_0^{-3} for a total charge $Z_{Co} = 0.405$ on the cobalt atom, a quadrupole coupling constant of $e^2qQ/h = -141.93$ MHz has been found. Although there is no quantitative agreement with the experimental value of -116.46 MHz, the order of magnitude and the sign are correct. It was found, however, that the orientation of the calculated quadrupole tensor does not agree with the experimentally determined directions. This discrepancy may be overcome if anisotropic contraction[160] of the cobalt 3 d in-plane orbitals, $(3\ d_{x^2-y^2}, 3\ d_{xy})$ is included $(\langle r^{-3}\rangle_{in\text{-}plane} = 6.3\ a_0^{-3})$. The contraction decreases the overlap with the ligand orbitals and leads to less cobalt-ligand covalency. This is in agreement with the fact that the experimental spin density on the nitrogen atoms[12, 59] is smaller than the calculated one[218].

Summing up, one can conclude that all the magnetic interactions found for Co(acacen) in Ni(acacen) · 1/2 H_2O are compatible with a $|^2A_2, yz\rangle$ ground state of this low-spin Co(II) compound.

[N,N′-ethylene-bis(salicylideniminato)]Co(II)pyridine, Co(salen)py

The first single crystal EPR study on a five-coordinated low-spin Co(II) *Schiff* base complex has been reported by Jörin et al.[80]. From the magnetic data of Co(salen)py diluted into a single crystal of Zn(salen)py, the orientation of the principal axes of the **g** and $\mathbf{A}^{Co}$ tensor with respect to the molecular structure could be determined, and the orientations of the in-plane **g** and hf tensor axes predicted by theory could be verified.

The hfs and quadrupole tensors of the axial pyridine nitrogen have been determined from ENDOR spectra of a Co(salen)py powder sample with $\mathbf{B}_0$ along g_x, g_y and g_z. The single crystal-like ENDOR spectrum with $\mathbf{B}_0$ along g_x is shown in Fig. 12. The magnetic parameters $A_1^N = A_2^N = 36.6$ MHz, $A_3^N = 44.4$ MHz and $Q_1^N = 1.08$ MHz, $Q_2^N = 0.48$ MHz, $Q_3^N = -1.59$ MHz with A_3^N and Q_3^N along g_z have been evaluated. Spin densities of $f_s^N = 0.025$ and $f_p^N = 0.054$ were found in the σ-bond of the pyridine nitrogen, resulting in a hybridization ratio $n^2 = 0.68$. The quadrupole coupling constant $e^2qQ/h = -3.16$ MHz is very close to the coupling constant of -3.27 MHz observed for the benzimidazole nitrogen in vitamin B_{12r}[228]. No hf coupling data of the equatorial nitrogen ligand nuclei in a five-coordinated low-spin Co(II) complex have been reported so far. From the single crystal EPR linewidth of Co(salen)py, an upper limit for this coupling of 6 MHz has been estimated[80].

Vitamin B_{12r}

Paramagnetic species of vitamin B_{12} have been extensively studied by EPR. The reported results of the reduced low-spin Cob(II)alamin (B_{12r}) and of its oxygenated form ($B_{12r}O_2$), evaluated from powder samples and frozen solutions, are often incomplete because of the poor resolution of the EPR spectra[229, 230]. A more detailed insight concerning the electronic and geometric structure of the biologically active center was made possible by studying B_{12r} and $B_{12r}O_2$ doped into B_{12b} (hydroxocobalamin) single crystals[231, 232].

A further improvement in resolution can be expected from ENDOR studies in frozen solutions or in single crystals. To determine the optimum ENDOR conditions regarding solvents and concentrations, various frozen B_{12r} solutions have been investigated near liquid helium temperature. For $\mathbf{B}_0$ along g_z (normal to the corrin plane) some poorly structured proton ENDOR spectra of B_{12r} in the *base off* (pH = 0) and *base on* (pH = 7) state with $|A^H| < 3$ MHz have been observed. For *base on* B_{12r} in H_2O/ethanol (ethanol is used as resolution enhancer[229]), the full hfs and quadrupole tensors of the benzimidazole nitrogen could be measured. The principal values normal to the corrin plane were determined from *difference spectra* of the signal obtained with $\mathbf{B}_0$ set at the fourth high-field triplet of the g_z features and *between* the third and the fourth high-field triplet. Assuming axially symmetric $\mathbf{A}^N$ and $\mathbf{Q}^N$ tensors, the coupling constant $A_\perp^N$ could be determined by saturating one of the extrema of the $g_\perp$-feature of the EPR spectrum. The magnetic nitrogen parameters obtained in this way were found to be $A_\parallel^N = 52.6$ MHz, $A_\perp^N = 47.5$ MHz and $Q_\parallel^N = -1.70$ MHz, $Q_\perp^N = -1/2 \cdot Q_\parallel^N = 0.85$ MHz. Due to the poor resolution of the frozen solution ENDOR spectra, deviations of $\mathbf{Q}^N$ from axiality, as observed in Co(salen)py, could not be measured in B_{12r}.

Nitrogen ENDOR spectra of frozen solutions of the copper-containing analogue of vitamin B_{12} have been used to verify the corresponding EPR data obtained from powder simulations[233]. The hf values from the very poorly resolved ENDOR spectra agree with the EPR results within experimental error. Since the errors estimated from the ENDOR spectra are only insignificantly smaller than those from the EPR powder spectrum, the ENDOR measurements will not alter the interpretation of the data.

6.3 Fe(III) Compounds (d^5)

Up to now, two types of iron compounds have been studied with ENDOR, namely heme compounds (hemoproteins and some heme model compounds) and iron-sulfur proteins. For comprehensive summaries of the corresponding EPR work, the reader is referred to the literature[234–237].

6.3.1 Hemes and Hemoproteins

EPR spectroscopy has been used very extensively in investigations of hemoproteins and heme model systems[237]. In all studies on high- and low-spin hemoproteins, no hf interaction with ^{14}N, ^{1}H or (enriched) ^{57}Fe was resolved in frozen solutions or in single crystals. The first EPR observation of ligand hf structure from heme pyrrole nitrogens in the model system ferriprotoporphyrin(IX) ester doped into perylene single crystals was reported by Scholes[238, 239]. The hfs tensors were found to be essentially isotropic with $|a_{iso}^N| = 8.68$ MHz, which corresponds to a spin density $f_s^N = 2.7\%$ in the nitrogen 2s-orbital.

The ENDOR literature on hemes and hemoproteins has recently been reviewed by Scholes[240)]. Papers which are discussed in Scholes' work will only briefly be summarized[7].

ENDOR signals from heme compounds, namely from (pure) metmyoglobin (Mb) and myoglobin azide single crystals were first reported by Eisenberger and Pershan in 1967[14)]. In Mb, only the proton matrix ENDOR line was observed, whereas in Mb-N_3^- some additional proton peaks from heme protons but no nitrogen signals could be seen. The poor quality of the ENDOR spectra was due to the fact that in pure Mb crystals the short heme-heme distance ($\simeq 25$ Å) produces strong spin-spin interactions which prevent the observation of nitrogen ENDOR.

High-Spin (S = 5/2) Metmyoglobin and Methemoglobin

Nitrogen ENDOR. The initial nitrogen ENDOR work on hemes and hemoproteins was published by Scholes et al.[241)], who studied aquo-myoglobin and hemin in frozen solutions. A representative nitrogen ENDOR spectrum of aquo-Mb with $\mathbf{B}_0$ along $g_{\|} = 2$ is shown in Fig. 44. Since all four heme nitrogens are approximately equivalent for this orientation, the single crystal-like ENDOR spectrum is easy to interpret. According to (3.10), four first order ENDOR transitions are expected for each of the two sets of nitrogens (heme nitrogens and proximal histidine nitrogen). The assignment of the ENDOR lines to the corresponding set of nitrogen nuclei was based on a comparison of the myoglobin spectrum with the four-line nitrogen ENDOR spectrum of hemin, which lacks the histidine nitrogen.

The magnetic parameters of aquo-Mb obtained from Fig. 44 are collected in Table 15.1. The theoretical A_3^N values for the heme and histidine nitrogen, which are about 50% smaller than the observed values, have been determined by Mun et al.[242)], using an extended Hückel-type calculation. According to these authors, the agreement between theoretical and experimental values could perhaps be improved further by considering electron core polarization effects.

In order to obtain the full hfs and quadrupole tensors of the nitrogen ligands, single crystals of diamagnetic CO-ligated ferrous Mb doped with about 10% high-spin ferric Mb have been prepared[240)]. Preliminary data of the principal values in the heme plane are

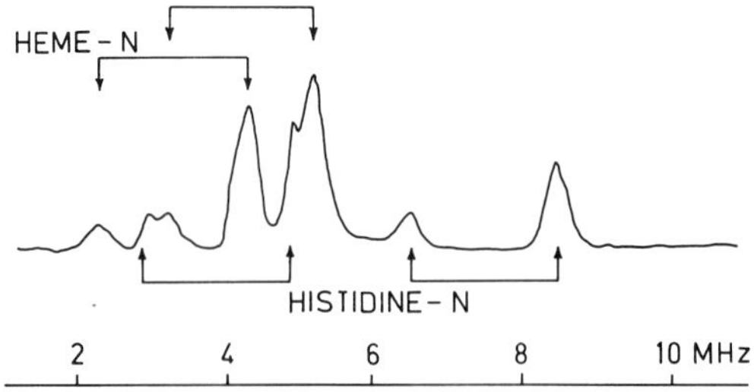

Fig. 44. Single crystal-like ^{14}N-ENDOR spectrum of metmyoglobin with $\mathbf{B}_0$ along $g_{\|}$; temperature 2.1 K (Mb chromatographically purified, 6 mM in 50% (v/v) glycerol, 0.1 M potassium phosphate buffer, pH = 6.0). (Adapted from Ref. 241)

7 It should be noted that the quadrupole coupling constants in the papers by Scholes and coworkers are given as $P_i(= 3/2\ Q_i)$

Table 15.1. Nitrogen coupling constants of aquo-myoglobin (data from Scholes et al.[240, 241] and Mun et al.[242]; in MHz)

Nitrogen ligand	Hf coupling		Quadrupole[a] coupling
	exp[b]	calc[c]	
heme[d]	$\|A_3\| = 7.60$	$A_3 = 3.68$	$\|Q_3\| = 0.29$
histidine[d]	$\|A_3\| = 11.46$	$A_3 = 8.18$	$\|Q_3\| = 1.16$
heme[e]	$\|A_1\| = 20.5$		$\|Q_1\| = 1.01$
	$\|A_2\| = 30.0$		$\|Q_2\| = 0.7$

[a] All the quadrupole data in the papers by Scholes and coworkers are given by P_i ($= 3/2\ Q_i$)
[b] From[240, 241]
[c] From[242]
[d] From frozen solutions at $g_{\|} = 2.0$ (perpendicular to the porphyrin plane)
[e] From dilute single crystals[240], preliminary values

Table 15.2. Proton hf couplings of aquo-myoglobin (data from Mulks et al.[245]; in MHz)[a]

Proton assignment	$\|A_3\|$[b]
heme bound H_2O	6.02
heme, meso proton	0.79
δ-N proton of prox. histidine	1.33

[a] Coupling constants for the corresponding fluoro compound are only slightly different
[b] From frozen solutions at $g_{\|} = 2.0$

given in Table 15.1. Unexpected splittings of the heme nitrogen ENDOR lines for $\mathbf{B}_0$ lying in the heme plane were reported, which might be traced back to noncrossing effects[67] similar to those discussed in Sect. 3.2.2.

The nitrogen ENDOR spectra of high-spin Mb-F^- and Hb-F^- are very similar to the spectrum of aquo-Mb[245]. The two high-frequency peaks of the histidine nitrogen described by $A_3^N/2 + 3/2\ Q_3^N \pm \nu_N$ are reported to be about 0.5 MHz lower in frequency than the corresponding transitions in the aquo-ligated hemoproteins. These shifts are traced back to a trans effect upon replacement of H_2O by an F^- ion.

An interesting application of ENDOR to hemoproteins has been reported on the mutant hemoglobins $M_{Hyde\ Park}(\alpha_2\beta_2^{92\mathrm{His}\rightarrow\mathrm{Tyr}})$ and $M_{Iwate}(\alpha_2^{87\mathrm{His}\rightarrow\mathrm{Tyr}}\beta_2)$, where the proximal histidines of the β and α chain, respectively, are replaced by tyrosine[240, 243]. The hemes in subunits with a replaced histidine are stabilized in the high-spin ferric form, which will not bind oxygen. Since tyrosine lacks a nitrogen for ligating to the heme iron, and the unaltered chains are in the diamagnetic ferrous form, only a four-line ENDOR spectrum has been observed for both mutants. The lack of histidine signals clearly indicates that the distal histidine has not taken over the role of the proximal histidine, in contrast to earlier speculations[244].

Some preliminary ENDOR investigations of the fascinating cooperative oxygenation effect in hemoglobin have been reported by Feher et al.[243]. Since magnetic resonance

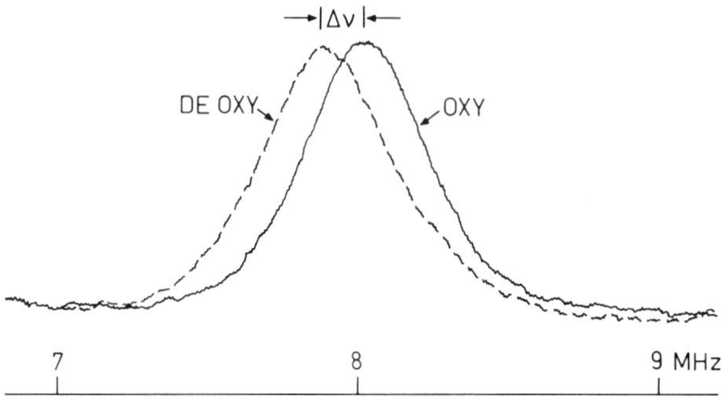

Fig. 45. Effect of oxygenation of Hemoglobin $M_{Milwaukee}$ ($\alpha_2\beta_2^{67\,Val \rightarrow Glu}$) on one of the histidine ^{14}N-ENDOR lines (Washed whole cells). (Adapted from Ref. 243)

involving electron spins requires heme subunits being in the ferric state, the authors made use of the mutant hemoglobin $M_{Milwaukee}$($\alpha_2\beta_2^{67\,Val \rightarrow Glu}$) for which both β subunits are in the ferric form. Upon oxygenation of the α subunits, the nitrogen ENDOR peak of the proximal histidine near 8 MHz shifts to higher frequencies by $\Delta\nu = 100$ kHz (Fig. 45). The observed shift clearly indicates that oxygenation of the heme in the α chains changes the environment of the ferric heme in the β chains.

Proton ENDOR. Frozen solution proton ENDOR spectra of aquo-Mb and Mb-F^- for $\mathbf{B}_0$ along $g_{\parallel} = 2.0$ were reported by Feher et al.[243] and by Mulks et al.[245]. The proton hf data of aquo-Mb are collected in Table 15.2. The corresponding Hb coupling constants are found to be essentially identical. The hf coupling constant $|A_3^H| = 6.02$ MHz observed in aquo-Mb is associated with the exchangeable protons of the water ligand and agrees well with the calculated dipolar coupling.

^{57}Fe-ENDOR. ^{57}Fe hf interactions in Mb and Hb have been reported by Scholes et al.[246]. The samples were enriched to about 90% in ^{57}Fe which has a nuclear spin of I = 1/2. Since the tensor $\mathbf{A}^{Fe}$ has to be (nearly) axially symmetric and coaxial with $\mathbf{g}$, single crystal-like ENDOR spectra may be recorded at *any* B_0 field position in the EPR spectrum; thus, the *complete* hfs tensor can be determined from a frozen solution sample. ENDOR frequencies with the (intrinsic) hf coupling parameters $A_{\parallel}^{Fe} = -27.77$ MHz (−27.15 MHz) and $A_{\perp}^{Fe} = -27.05$ MHz (−26.70 MHz) have been found for Mb(Hb).

Low-Spin (S = 1/2) Metmyoglobin and Methemoglobin

ENDOR on low-spin Mb and Hb derivatives in frozen solutions has been studied not only at g_z (which points approximately along the complex normal) but also at the medium value g_y[76] and at g_x[245].

Nitrogen ENDOR. For $\mathbf{B}_0$ along g_z and g_x[245], the hemoproteins Mb-N_3^- and Hb-N_3^- show well resolved and intense nitrogen ENDOR signals. The spectra, however, are difficult to analyze because of their complicated structure and the considerable overlap of the nitrogen ENDOR lines. For the assignment of the transitions to particular groups of nitrogens, the authors made use of samples enriched in ^{15}N (^{15}N-heme or ^{15}N-azide), which has a nuclear spin I = 1/2 and generates only a two-line ENDOR pattern.

The analysis shows that all the observed transitions can be assigned to the four heme nitrogens which are *no longer equivalent* for $\mathbf{B}_0$ along g_z. The magnetic hf parameters $|^{15}A_3^N|$ are in the range 7.0–8.5 MHz, while $|^{14}A_3^N|$ and $|Q_3^N|$ are in the range 5.0–6.5 MHz and 0.25–0.4 MHz, respectively. The ^{15}N data do not agree well with the ^{14}N parameters, probably because second order ^{14}N quadrupole shifts (3.14) affect the ENDOR frequencies. This is particularly true for the low-frequency ENDOR transitions. From the ability to follow resolved ^{15}N lines for $\mathbf{B}_0$ fields between g_z and g_x, the isotropic coupling was found to be the dominant contribution to the transition frequencies. The strong peaks observed in the 7.0–8.0 MHz range for $\mathbf{B}_0$ along g_x were assigned to $\Delta m_I = \pm 2$ transitions[70].

^{13}C-ENDOR. ^{13}C-ENDOR has been observed in Mb-CN^- and Hb-CN^- samples prepared with ^{13}C enriched $K^{13}CN$[245]. For $\mathbf{B}_0$ along g_z, ^{13}C hf coupling constants of 28.64 MHz (Mb-$^{13}CN^-$) and 27.33 MHz (Hb-$^{13}CN^-$) are reported. ENDOR spectra obtained at arbitrary g-values between 2.0 and 3.4 indicate that the ^{13}C hf coupling is dominantly isotropic. From comparison with NMR and EPR data of other transition metal cyanide complexes it is assumed that the sign of the isotropic coupling will be negative.

Proton ENDOR. A few proton ENDOR data have been reported on Mb-CN^-, Hb-CN^-[245] and Mb-N_3^-[76]. In hemoprotein cyanides with $\mathbf{B}_0$ along g_z, the largest hf coupling constant $|A_3^H| \simeq 4$ MHz is tentatively assigned to the near-CH-protons of the proximal histidine. A signal of an exchangeable proton with $|A_3^H| = 1.75$ MHz originates most likely from the proton on the δ-nitrogen of the proximal histidine or from a proton that is hydrogen-bonded to the cyanide. The hf values $|A_3^H| = 1.34$ MHz (Hb-CN^-) and $|A_3^H| =$ 1.19 MHz (Mb-CN^-) are assigned to the meso protons.

An interesting interpretation of the proton ENDOR data obtained in Mb-N_3^- for $\mathbf{B}_0$ along g_y has been given by Rist[76]. Although not only molecules with g_y along $\mathbf{B}_0$ are saturated for a magnetic field setting at g_y, a well-resolved proton ENDOR spectrum has been observed. From EPR on Mb-N_3^- single crystals, the angle between g_y and the direction of one pair of (nearly) equivalent meso protons is known to $\varphi = 16°$ [248]. Rist[76] has found that for an angle $\varphi = 19°$ the splitting of the two sets of observed proton peaks with $|A_2^H| = 1.65$ MHz and 0.71 MHz is in excellent agreement with calculated dipolar couplings of the meso protons. This result demonstrates that the orientation of the **g** tensor principal axes in the heme plane may be determined by ENDOR from a *powder* sample.

Cytochrome P 450$_{CAM}$

The hemoprotein cytochrome P 450$_{CAM}$ is found in the bacterium *Pseudomonas putida* and catalyzes the 5-hydroxylation of the substrate D-(+)-camphor. There is some controversy in the literature over the axial ligation at the heme-iron active site of the substrate free, low-spin ferric m^o form. One axial ligand is supposed to be a cysteinyl sulfur. From pulsed EPR measurements the second ligand is suggested to be imidazole[249], whereas pulsed NMR results imply the presence of strongly coupled exchangeable protons within 2.6–2.9 Å of the m^o heme iron[250]. When substrate is added, most of the low-spin m^o form is converted to the high-spin m^{os} form which exhibits unusually highly anisotropic g-values $g_x = 8$, $g_y = 4$ and $g_z = 1.8$[251].

Information about the axial ligands in both the low-spin m^o and the high-spin m^{os} form has been provided from ENDOR by LoBrutto et al.[252].

Low-spin m^o form. Besides some poorly resolved nitrogen ENDOR transitions below 6 MHz, a strongly coupled exchangeable proton with $|A_3^H| = |a_{iso}^H + A_d^H| = 8$–$9$ MHz has been observed for $\mathbf{B}_0$ along g_z. a_{iso}^H is estimated to 2.2–3.1 MHz from proton relaxation measurements[253]. The remaining dipolar coupling A_d^H corresponds to protons 2.6–2.9 Å away from the heme plane along g_z. According to LoBrutto et al.[252] these protons could belong to H_2O, RNH_2, ROH or $RCONH_2$, or be exchangeable protons bound to cysteinyl sulfur.

High-spin m^{os} form. A comparison of the ENDOR spectrum of aquo-Mb with that of cytochrome P 450$_{CAM}$ in the high-spin m^{os} form at $g_z = 1.8$ indicates that no resonances attributable to histidine nitrogen or to any exchangeable proton are present in the latter. Scholes and coworkers[79] have found that the coupling of the meso protons in high-spin heme compounds are greater in five- (0.95–1.01 MHz range) than in six-coordinated (0.79–0.83 MHz range) systems. In m^{os}, a hf value of $|A_3^H| = 0.93$ MHz for the meso protons is found, which is very close to the coupling of a five-coordinated high-spin ferric compound. From this argument it is inferred that the heme iron in the high-spin m^{os} form is five-coordinated.

The paper of LoBrutto et al.[252] is an interesting example of how the large body of ENDOR data available in heme compounds may be applied to interpret ENDOR spectra of heme systems with less known structure.

NO-Ligated Cytochrome *c* Oxidase

Reaction of nitric oxide with ferrohemoproteins produces paramagnetic NO-ligated heme proteins ($S = 1/2$, rhombic $\mathbf{g}$ tensors with principal values in the range 1.96–2.08). In many compounds studied so far by EPR the hf interaction of the NO nitrogen and of a second axial nitrogen is clearly resolved in the intermediate g-value region near $g = 2.0$[237].

Recently, King et al.[254] have shown that the nitrogen coupling of NO-ligated cytochrome *c* oxidase may be resolved at g_{min} and g_{max} using the ENDOR technique. For $\mathbf{B}_0$ along $g_{max} = g_x = 2.08$, the couplings $|A_1^N| = 30.6$ MHz and $|Q_1^N| = 1.1$ MHz have been reported. The authors believe that the good resolution of the quadrupole coupling for this orientation is an indication that g_{max} is directed along a line in the heme plane which is perpendicular to the Fe-NO plane.

High-Spin Heme Model Compounds

To get a deeper insight into the factors which affect the electronic structure of hemoproteins, systematic studies on heme model compounds have been undertaken by Scholes and coworkers[79, 247, 255]. The model systems used by these authors are protohemin, deuterohemin and ferric tetraphenylporphyrin with a series of axial ligands. All the published ENDOR data on high-spin heme model compounds were obtained in frozen solutions with $\mathbf{B}_0$ along $g_\parallel = 2$ which is oriented normal to the porphyrin plane. A 1:1

(v/v) mixture of THF-$CHCl_3$ (solvent I) or a 1 : 1 (v/v) mixture of CH_2Cl_2-$CHCl_3$ (solvent II) containing diamagnetic metal-free mesoporphyrin were used as solvents. These solvents do not replace the axial ligand of the metal complex, form glassy matrices on freezing, and prevent aggregation of the iron complexes.

The EPR signals of the model compounds with the axial ligand fluoride, acetate, formate, azide, chloride or bromide are typical high-spin spectra with $g_\perp$-values in the 5.8 to 5.9 range[79]. Splittings at $g_\parallel = 2$ due to the hf interaction of these ligands are only observed for fluoro- and bromo-hemins.

Nitrogen ENDOR. In all high-spin heme derivatives a four-line nitrogen ENDOR pattern has been observed for $\mathbf{B}_0$ along $g_\parallel$, indicating that the heme nitrogens are equivalent for this particular orientation. The magnetic parameters of the heme nitrogens using solvent I are collected in Table 16.1. A_3^N and Q_3^N are found to be quite insensitive to variations of the axial anion. Little change in $|A_3^N|$ with axial halides has also been found by Mallick et al.[256] in a recent extended Hückel-calculation on iron porphyrin. The theoretical values for fluoro-, chloro- and bromohemin are all about 50% of the experimental data.

A pronounced upward shift of $\Delta A_3^N = 0.3$–0.7 MHz has been reported upon going from solvent I to solvent II. It is striking that the smaller nitrogen hf coupling constants in THF-$CHCl_3$ are similar to those found in aquo-Mb and aquo-Hb, which are assumed to

Table 16.1 ^{14}N hf and quadrupole coupling constants of protohemin DME near $g_\parallel$[a]. Solution: THF-$CHCl_3$ (data from van Camp et al.[79]) and Mallick et al.[256]; in MHz)

Anion	$\|A_3\|$[b]	A_3[c]	$\|Q_3\|$[b]
F^-	7.54	3.67	0.24
$HCOO^-$	7.28		0.25
AC^-	7.48		0.25
N_3^-	7.45		0.21
Cl^-	7.41	3.80	0.25
Br^-	7.21	3.66	0.25

[a] ENDOR measured at fields 1 mT above the peak of the $g_\parallel = 2.0$ extremum (see Sect. 4.1)
[b] ENDOR data[79], [c] Theoretical values[256]

Table 16.2. Hyperfine coupling constants from axial F^-, Cl^- and Br^- ligands of protohemin DME for $\mathbf{B}_0$ along $g_\parallel$[a] (data from van Camp et al.[79]; in MHz)

Anion	In THF-$CHCl_3$		In $CHCl_3$-CH_2Cl_2 + mesoporphyrin	
	$\|A_3\|$	$\|Q_3\|$	$\|A_3\|$	$\|Q_3\|$
F^- [b]	132.4		128.6	
$^{35}Cl^-$ [c]	16.04	2.71	14.78	2.89
$^{37}Cl^-$ [c, d]	13.34	2.13	12.28	2.30
Br^- [b]	76.7		72.4	

[a] ENDOR resonances in the 5–10 MHz region in azidohemin have been assigned to the axial nitrogen
[b] EPR, [c] ENDOR, [d] 90% ^{37}Cl-enriched

be six-coordinated. In CH_2Cl_2-$CHCl_3$, on the other hand, the hf coupling constants are in the same range as those of five-coordinated hemin systems. It is assumed[79] that THF is capable of coordinating with the ferric ion via the oxygen. Hemins in solvent I may then serve as models for six-coordinated hemoproteins, whereas in solvent II the hemin is assumed to be five-coordinated.

Hf interaction with halides. The hf couplings of the axial halides are given in Table 16.2. Data for ^{35}Cl and ^{37}Cl are determined from ENDOR frequencies using Eq. (3.10). The values of $|A_3^{Cl}|$ are lower in solvent II than in solvent I. The degree of covalent bonding between the anion and the iron should increase upon going from F^- to Cl^- to Br^- (decreasing electronegativity). Since only the hf components $|A_3|$ are known, contact and dipolar contributions cannot be separated. According to van Camp et al.[79] the ratio of $|A_3|/g_n$ may be used to estimate covalency changes from one halide to the next. As expected, the values of $|A_3|/g_n$ are found to be in the order $F^- < Cl^- < Br^-$.

Low-Spin Heme Model Compounds

Only a few ENDOR data have been reported on low-spin heme model compounds[247]. In hemin mercaptide a four-line nitrogen ENDOR pattern with $|A_3^N| = 5.42$ MHz and $|Q_3^N| = 0.24$ MHz has been observed at $g_{max} = 2.32$, which is known to lie along the heme normal. Since this complex lacks the 5th and 6th nitrogeneous ligand, the resonances can be unambiguously assigned to heme nitrogens. It was hoped that the above magnetic parameters could be used for a definitive assignment of the nitrogen peaks to heme nitrogen or to histidine nitrogen in low-spin hemoproteins. Unfortunately, the ^{14}N spectrum of the mercaptide derivative does not sufficiently resemble the spectra of the hemoproteins, so that this assignment problem had to be solved using ^{15}N labeled heme.

The ENDOR work on heme compounds is an impressive example of a systematic study on an important class of biological compounds. One disadvantage is that most experiments had to be performed with the $\mathbf{B}_0$ field parallel to the complex normals, so that the full hfs and quadrupole tensors could usually not be evaluated. A single crystal ENDOR study on aquo-Mb is in preparation to rectify this lack of information[257].

6.3.2 Iron-Sulfur Proteins

The state of knowledge on iron-sulfur proteins which contain non-heme iron bonded to sulfur ligands (cysteinyl residues from the protein and inorganic sulfur) has been reviewed by several authors[258–264]. With magnetic resonance techniques it has been possible to obtain detailed information on the nature of the active site in many of these proteins. The contributions from ENDOR have recently been summarized by Sands[265] so that we shall only give an outline of the crucial points.

Besides a poorly resolved triplet structure in the EPR spectrum of ^{57}Fe enriched two-iron ferredoxins[266] at g_z, no hf interactions are observed in all these iron-sulfur clusters. The distribution of principal axes of the **g** tensors gives rise to an inhomogeneous, field-dependent EPR linewidth, and obscures the resolution of ligand hfs[265]. This distribution, whose origin is not completely established, does not, however, prevent application of the ENDOR technique.

Two-Iron Ferredoxins

The most likely structure of the two-iron-sulfur proteins is shown schematically in Fig. 46. The EPR spectrum of reduced 2-Fe ferredoxin below 77 K is described by a net spin S = 1/2 resulting from the antiferromagnetic exchange interaction between a high-spin ferric and a high-spin ferrous ion. ^{57}Fe hf data have been evaluated by comparing frozen solution ENDOR spectra of the native and the ^{57}Fe enriched protein[268, 269]. The resonances of one of the ^{57}Fe atoms lie above 20 MHz and are well separated from the proton ENDOR lines. The nuclear transitions of the second ^{57}Fe are located under the proton signals. For adrenodoxin and putidaredoxin the magnetic parameters of this second iron atom were obtained from ^{57}Fe-^{56}Fe difference spectra[268]. In the case of spinach ferredoxin and parsley ferredoxin, however, the A_1^{Fe} and A_2^{Fe} components were lost under the strong proton ENDOR signal and only the A_3^{Fe} value could be measured. For determination of the full hfs tensor of the smaller ^{57}Fe coupling, a completely deuterated ferredoxin, isolated from *S. lividus*, which is nearly identical with the spinach ferredoxin with respect to molecular weight, EPR, ENDOR and optical spectra has been prepared[269]. The ^{57}Fe-ENDOR data listed in Table 17 demonstrate that two different iron sites are present in the 2-Fe ferredoxins. These data proved to be useful in the

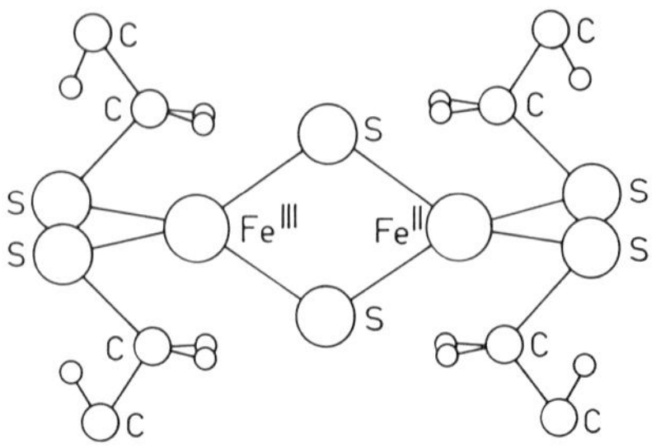

Fig. 46. Schematic structure of the iron-sulfur complex in two-iron ferredoxin and adrenodoxin. (Adapted from Ref. 267)

Table 17. ^{57}Fe hf principal values of two-iron ferredoxins (data from Fritz et al.[268] and Anderson et al.[269]; in MHz)

Protein	Ferric site			Ferrous site		
	A_1	A_2	A_3	A_1	A_2	A_3
Spinach ferredoxin[a]	51.0	49.9	41.9	–	–	35.6
Parsley ferredoxin[a]	50.7	49.9	41.9	–	–	34.7
Adrenodoxin[a)]	48.8	56.7	43.6	16.6	24.4	35.6
Putida-redoxin[a]	48.8	56.7	43.6	16.6	24.4	35.6
S. Lividus ferredoxin[b]	51.5	49.9	41.9	12.9	14.8	36.5

[a] From[268] [b] From[269]

simulation of the Mössbauer spectra of the reduced ferredoxins, which on the other hand permitted the assignment of the two ^{57}Fe hfs tensors to two antiferromagnetically coupled ferrous (S = 2) and ferric (S = 5/2) atoms.

A number of poorly resolved proton ENDOR peaks have been observed between 10–20 MHz[269–271]. Only small changes of the overall shape of the ENDOR spectrum were detected when the sample was freeze-dried and redissolved in D_2O, i.e. no strongly coupled exchangeable protons were present. From comparison with the proton ENDOR spectrum in an anhydrous powder, it was assumed that the signals arose from the methylene protons of the cysteine ligands and that the iron-sulfur chromophore was not exposed to solvent water[270].

In ferredoxins the labile sulfur atoms can be substituted by selenium without affecting the biological activity of the protein[271]. The observed hf couplings of Fe(III) are slightly reduced in selena-adrenodoxin (5–10%), indicating that the spin density on the active site is only weakly changed by selenium substitution. No evidence of any ^{77}Se-ENDOR signal (I = 1/2 nucleus) was found between 12 and 28 MHz.

Four-Iron-Sulfur Proteins

The 4-Fe proteins show a cubane-like iron-sulfur center with an iron atom at each of the four alternate corners of a distorted cube. The oxidized high-potential 4-Fe cluster is paramagnetic, with $S_{eff} = 1/2$, and displays a nearly axial EPR spectrum. A K_u-band ENDOR study of such an iron protein isolated from *C. vinosum* (^{57}Fe enriched) has been reported by Reid[272]. In this microwave frequency region the proton ENDOR signals do not overlap with the ^{57}Fe resonances. The hfs of ^{57}Fe were found to be nearly isotropic with coupling values of 22.6 and 32.5 MHz. Similar values have been found by Anderson et al.[269] from ^{57}Fe–^{56}Fe difference spectra. Since only *two* different $\mathbf{A}^{Fe}$ tensors were observed, the iron atoms were classified in two pairs of equivalent nuclei. Mössbauer data show that the two hf couplings are of opposite sign[265].

Eight-Iron-Sulfur Proteins

The 8-Fe proteins contain two four-iron-sulfur cubes which are separated (center-to-center) by about 12 Å. The EPR spectra from the fully reduced proteins are typical for a 4-Fe center with intercluster spin-spin coupling. An ENDOR study on the 8-Fe ferredoxin from *C. pasteurianum* in the fully reduced state has been reported by Anderson et al.[273]. Since the ^{57}Fe–^{56}Fe difference spectra are only poorly resolved, analysis of the ENDOR data turned out to be difficult. All eight iron atoms are assumed to have similar $\mathbf{A}^{Fe}$ tensors with principal values $A_1^{Fe} = 25$ MHz, $A_2^{Fe} = 29$ MHz and $A_3^{Fe} = 33$ MHz. These coupling parameters suggest that the reducing electrons are delocalized over the four irons of each of the two nearly identical 4-Fe clusters.

The relevant structural and chemical information on iron-sulfur proteins provided by EPR, ENDOR and Mössbauer techniques serves as an instructive example for the success of combining different types of spectroscopic methods.

6.4 V(0) Compounds (d^5)

Vanadium Dibenzene $V(bz)_2$

In most paramagnetic sandwich complexes the proton hf structure is not resolved in low temperature powder or single crystal EPR spectra. In vanadium dibenzene, a typical example of this type of compound, the poor resolution is due to the fact that the aromatic rings are rigidly frozen at $T < 50$ K and thus the proton hfs tensors of the benzene rings are no longer magnetically equivalent.

$V(bz)_2$ has been extensively studied in different solvents by EPR and NMR[274–276]. The isotropic coupling a^H_{iso} of the benzene protons obtained from solution EPR is found to be rather insensitive to solvent effects, as expected for a $^2A_{1g}$ ground state. The room temperature values A^H_z (along the rotation axis) and A^H_{av} (normal to this axis) of the partially averaged proton tensors (quasi-free rotating benzene rings) have been determined from EPR spectra of $V(bz)_2$ in $Fe(cp)_2$ and $Cr(bz)_2$ host crystals[274].

To evaluate the full proton hfs tensors for the rigid molecule and to study the temperature dependence of the dynamics of the ring rotation, ENDOR and EI-EPR spectroscopy has been applied to powder samples of these two systems[37, 78]. EPR, ENDOR and EI-EPR data of $V(bz)_2$ diluted into $Fe(cp)_2$ are summarized in Table 18.

Rigid molecules. From the two-dimensional powder ENDOR spectrum (Sect. 4.1) at 15 K with $\mathbf{B}_0$ in the benzene ring plane (EPR observer marked by an arrow in Fig. 15a), the intermediate principal value A^H_1 (smallest splitting) and $A^{H'}_2$ (largest splitting), which is *not* a principal value of the proton tensor, could be determined. In the system $V(bz)_2 : Fe(cp)_2$ the two principal values A^H_2 and A^H_3 and their principal directions have been measured using EI-EPR[37] (see Fig. 15). From these data the interatomic V-H distance and the angle between the principal axis A^H_3 and the axis normal to the benzene rings are calculated to be 2.96 Å (point-dipole model) and 56°, respectively. Comparison with values of 2.94 Å and 57° from neutron diffraction data of $Cr(bz)_2$[277] shows that the point-dipole model with one unpaired electron at the vanadium nucleus is a good approximation. The rhombic distortion of the hfs tensor (Table 18) can be explained by a small spin density on the adjacent carbon atom.

Table 18. Proton hf data of $V(bz)_2$ diluted into $Fe(cp)_2$; in MHz

Hf splittings[a]	EPR[b]	ENDOR[c]	EI-EPR[d]
A_z ($\parallel g_\parallel$)	10.6		
A_1 ($\parallel g_\perp$)		9.2	
A_2			7.4
A'_2 ($\parallel g_\perp$)		14.4	
A_3		$\simeq 17.0$	17.5
A_{av}	11.4	11.8	
a_{iso}	11.1		

[a] A_1, A_2, A_3: principal values
[b] A_z, A_{av}: single crystal data, $T = 290$ K, from[274]
a_{iso}: liquid solutions, $T = 290$ K, from[274, 275]
[c] A_1, A'_2, A_3: $T = 15$ K, A_{av}: $T = 163$ K, from[78]
[d] $T = 15$ K, from[37]

Dynamics of ring rotation. In both host compounds, $Fe(cp)_2$ and $Cr(bz)_2$, the rotation of the benzene rings of $V(bz)_2$ has been studied by ENDOR in the temperature range $T = 15 - 180\,K$[78]. The temperature dependence of the two-dimensional powder ENDOR spectrum of $V(bz)_2 : Fe(cp)_2$ is shown in Fig. 47 (high-frequency region). For $T < 70\,K$ the rotation frequency is slow compared to the time scale of the ENDOR experiment, i.e. the ENDOR signal corresponds to the spectrum of the rigid molecule and is limited by the frequencies $\nu_p + A_1^H/2$ and $\nu_p + A_2^{H'}/2$. With increasing temperature the rotation of the benzene rings narrows the spectrum, which coalesces for $T > 160\,K$ into a single line at the average ENDOR frequency $\nu_p + (A_1^H + A_2^{H'})/4 = \nu_p + A_{av}^H/2$. Closer inspection of the spectra shows that the dynamic behavior of the benzene rings of $V(bz)_2$ in $Fe(cp)_2$ occurs in two steps, with transition temperatures $T_{c1} = 110 \pm 10\,K$ and $T_{c2} = 145 \pm 10\,K$, in contrast to the system $V(bz)_2 : Cr(bz)_2$, where a single transition temperature with $T_c = 150 \pm 20\,K$ has been found[78]. The two transition temperatures in $V(bz)_2 : Fe(cp)_2$ can be explained in terms of at least two different barriers of internal rotation due to different packing forces. This is in agreement with the low temperature

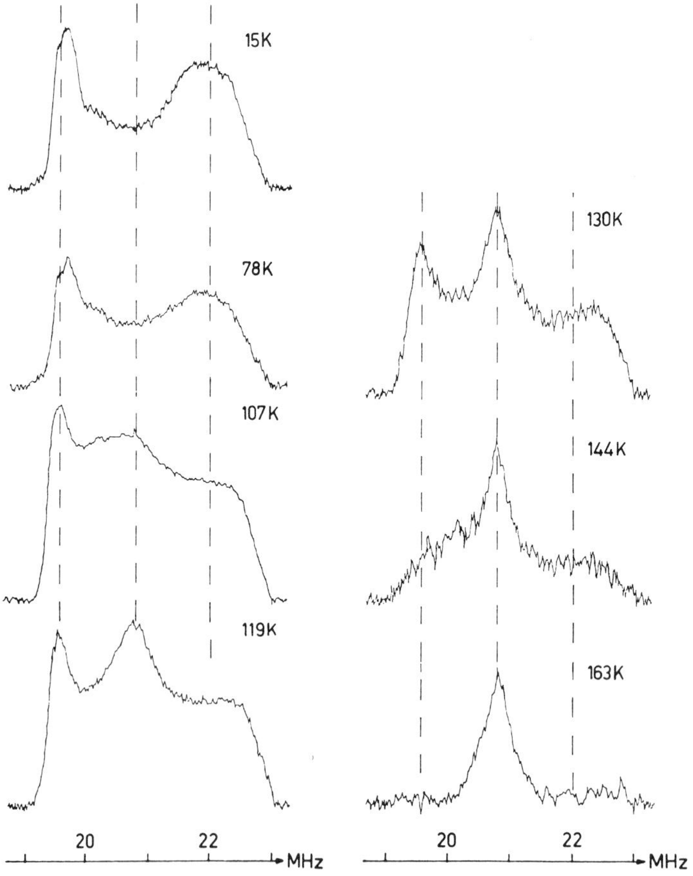

Fig. 47. Temperature dependence of the high field part of the two-dimensional proton ENDOR spectrum of $V(bz)_2$ diluted into ferrocene. (From Ref. 78)

X-ray structure of $Fe(cp)_2$ which contains two crystallographically nonequivalent $Fe(cp)_2$ molecules with slightly different atomic distances and different packings[278, 279].

The ENDOR data demonstrate that intermolecular interactions dominate the potential function for the ring rotation. The fast exchange limit of $V(bz)_2$ is reached at lower temperatures in $Fe(cp)_2$ than in $Cr(bz)_2$. This is in qualitative agreement with the NMR relaxation results of Campbell et al.[280] on the pure host materials. Since a transition temperature of $T_c = 79$ K is predicted for $Fe(cp)_2$ from the NMR data, one can conclude that the dynamic behavior of the guest molecules is not entirely determined by the host properties alone, but that the guest $V(bz)_2$ introduces a significant local perturbation into the host lattice by its larger size.

A single crystal ENDOR study on $V(bz)_2$ diluted into 2,2-paracyclophane is in preparation, which will allow to get deeper insight into the dynamic behavior of the benzene rings in these types of compounds[281].

6.5 Ti(III) Compounds (d^1)

Ti(η-C_8H_8) (η-C_5H_5)

Proton ENDOR and DOUBLE ENDOR spectra of a frozen solution of the mixed sandwich compound Ti(η-C_8H_8) (η-C_5H_5) have been reported by Labauze et al.[282]. The shape of the powder EPR spectrum at T = 150 K was found to depend strongly upon the rate of freezing and the nature of the solvent. Rapid freezing in toluene, for example, yields a glass with randomly trapped paramagnetic centers (axial **g** tensor: $g_\parallel = 2.000$, $g_\perp = 1.971$). On the other hand, in rapidly frozen THF or dichloro-methane the complex is free to rotate around an axis *parallel* to the ring planes and an averaged **g** tensor with $'g_\perp' = (g_\parallel + g_\perp)/2 = 1.988$ and $'g_\parallel' = 1.971$ is found. In slowly cooled toluene or THF a mixture of these two spectra is observed.

The proton ENDOR spectra recorded at different g-turning points clearly demonstrate that, irrespective of the solvent and the freezing procedure, the two aromatic rings are *quasi-free rotating* at T = 150 K (fast exchange limit), as already found for $V(bz)_2$. Unfortunately, the freezing of this ring rotation at T < 150 K was not studied. The proton hf parameters for T = 150 K as well as the calculated spin densities and dipolar coupling constants are collected in Table 19.

Table 19. Proton hf coupling constants of Ti(η-C_8H_8) (η-C_5H_5) (data from Labauze et al.[282]; in MHz)

	$C_8H_8^{2-}$	$C_5H_5^-$
$'A_\parallel'^a$	7.70	4.95
$'A_\perp'^a$	9.52	3.76
$a_{iso} = ('A_\parallel' + 2'A_\perp')/3$	8.91	4.15
s-character	0.62%	0.29%
$A_{dipolar} = ('A_\parallel' - a_{iso})$	−1.21	0.80
$A_{dipolar}$ (calc)	−1.04	0.42

[a] *Not* a principal value of the hfs tensor

For an assignment of the observed hf couplings to the corresponding protons in the two ring systems, DOUBLE ENDOR (special TRIPLE,Sect. 4.3) has been applied. The intensity ratio of the two DOUBLE ENDOR proton transitions is about 8 : 5, this confirms the earlier EPR observation[283] that the larger hf coupling arises from the C_8H_8 ring. It should be mentioned, however, that in such a DOUBLE ENDOR experiment the intensities are proportional to the number of participating equivalent nuclei only, if cross relaxation effects can be ignored[90]. This is usually not the case for transition metal complexes, so that application of this technique might be questionable. The *relative signs* of the proton hf couplings have been evaluated according to the DOUBLE ENDOR scheme in Fig. 16.

The spin densities deduced from the formula $a^H_{iso} = ('A^{H\prime}_{\parallel} + 2'A^{H\prime}_{\perp})/3$ are presumed to be positive because of direct delocalization of the unpaired electron from the $(a\,3d_{zz} + b\,4s)$ hybrid orbital. It should be noted, however, that $'A^{H\prime}_{\parallel}$ and $'A^{H\prime}_{\perp}$, measured for rotating rings at $T = 150\,K$, are *not* principal values of the proton hfs tensor, i.e. a^H_{iso} determined in this way does not exactly correspond to the isotropic coupling obtained by the common formula $a^H_{iso} = (A^H_1 + A^H_2 + A^H_3)/3$.

The small spin density at the proton in this titanium complex is close to the values found for other sandwich compounds ($V(C_7H_7)(C_5H_5)$: 0.89% and 0.35%[284], $V(bz)_2$: 0.80%[78], $Cr(bz)_2^+$: 0.68%[285]).

Similar to that found in $V(bz)_2$, the true proton hfs tensor in $Ti(C_8H_8)(C_5H_5)$ is expected to be rhombically distorted. In such a case only the value of the principal axis which points along the electron-proton direction should be used for determination of interatomic distances[62, 78]. We believe that the discrepancy between calculated and measured dipolar couplings in Table 19 may be due to this nonaxiality of the tensors rather than to second order effects. Although the Ti-H distances are almost identical for both rings, the hf coupling constant of the C_8H_8 protons is about twice that of the C_5H_5 protons. This because the C_5H_5 protons are closer to the nodal cone of the d_{zz} orbital, so that a smaller spin density is expected.

The ENDOR data obtained from samples with mixed EPR spectra provide further evidence that in the various types of cavities or cages in certain frozen solvents a rotation of the titanium sandwich compound around an axis parallel to the ring planes becomes possible.

6.6 VO(II) Compounds (d^1)

The vanadyl ion VO(II) is extensively used in EPR studies of organic and inorganic compounds as well as in metal ion binding site structures in systems of biological importance. For a thorough discussion of the EPR investigations the reader is referred to the reviews of Kokin[286] and Chasteen[287]. Since the ligand hf structure is usually not resolved in the EPR spectra of VO(II)-complexes[162], ENDOR spectroscopy proved to be a suitable technique to study these interactions[56, 60, 71, 288–290].

ENDOR on VO(II) complexes is facilitated for several reasons[289]: (a) the EPR transitions can easily be saturated (ENDOR signals are often observed at 100 K), (b) the large anisotropy of the $\mathbf{A}^V$ tensor allows for a high orientation selectivity in powder samples, and (c) the V=O bond may be used as internal reference axis. Moreover, the $\mathbf{g}$ tensor is nearly isotropic so that contributions to the hf interactions from an unquenched orbital moment may be neglected (Sect. 5.1).

Vanadyl Tetraphenylporphyrin, VO(TPP)

Proton and nitrogen ENDOR results on VO(TPP) diluted into Zn(TPP) powder, and in $CHCl_3$-toluene and pyridine frozen solutions have been reported by Mulks and van Willigen[71]. The well-resolved single crystal-like ENDOR spectra along the V=O axis could easily be interpreted using Eq. (3.10). Data from the two-dimensional nitrogen ENDOR spectrum with $\mathbf{B}_0$ in the complex plane, however, had to be evaluated using computer simulations. For the system VO(TPP): Zn(TPP) the following nitrogen hf and quadrupole parameters have been found (a) in the porphyrin plane: $A_1^N = 9.57$ MHz, $Q_1^N = \pm 0.16$ MHz, $A_2^N = 2.90$ MHz, $Q_2^N = \mp 0.41$ MHz (b) normal to the porphyrin plane: $A_3^N = 7.92$ MHz, $Q_3^N = \pm 0.25$ MHz. The relative sign of the hf values have been evaluated from computer simulations, which show pronounced changes of the ENDOR intensities for different sign combinations.

In VO(TPP) the unpaired electron in the $3d_{xy}$ orbital has a minimal overlap with ligand orbitals since the lobes of d_{xy} bisect the N-V-N bond angles. In-plane π-bonding and admixture of $\sigma(d_{x^2-y^2})$ character in the MO of the unpaired electron could be responsible for the observed ^{14}N contact interaction of 6.8 MHz[71]. Since axial coordination of solvent molecules does not affect the in-plane d_{xy} orbital, no striking changes in the ENDOR spectrum will be expected. Indeed, the ^{14}N data for VO(TPP) in Zn(TPP) or in frozen solutions of $CHCl_3$-toluene or pyridine are found to be very similar. The weak ENDOR peaks below 2 MHz observed in VO(TPP): pyridine may be attributed to the coordinated pyridine nitrogen.

In VO(TPP): $CHCl_3$-toluene the measured hf values of the pyrrole protons are found to be 1.38 MHz and −0.26 MHz (in-plane) and −0.26 MHz (out-of-plane). The hfs tensor may be separated into a contact contribution $a_{iso}^H = 0.29$ MHz and an axial dipolar contribution $A_\parallel^H = 1.10$ MHz which corresponds to a V-H distance of 5.24 Å. The axiality of the dipolar tensor indicates that the spin density on the nitrogen ligands is very small in VO(TPP) (in agreement with the ^{14}N hfs data), in contrast to Ag(TPP), where the unpaired electron is significantly delocalized and strongly rhombic pyrrole proton tensors have been observed[66]. No interpretation of the ^{14}N quadrupole interaction has been given. As found in $VO(acac)_2$, the proton resonances in VO(TPP) show a considerable solvent dependence. Since interpretation of this effect is highly speculative at present, we shall not discuss this problem.

EPR and ENDOR spectra very similar to those found in VO(TPP) have also been observed[288] in a sample of a black solid obtained by chloroform extraction of a phosphate mineral from Youssoufia (Morocco). The close correspondence of the spectra suggests that the black solid contains a vanadyl porphyrin derivative.

Vanadyl-bis(2,4-pentanedionate), $VO(acac)_2$

Proton ENDOR and DOUBLE ENDOR data on $VO(acac)_2$ in frozen solutions have been reported by van Willigen[289]. Since the CH-protons exchange with deuterons in CD_3OD they may easily be discriminated from the methyl protons. From the single crystal-like ENDOR spectrum for $\mathbf{B}_0$ along $g_\parallel$ and the two-dimensional powder spectrum at $g_\perp$, the hf coupling constants $A_2^{CH_3} = 0.7$ MHz and $A_1^{CH_3} = 0.7$ MHz, $A_3^{CH_3} = 1.5$ MHz, respectively, have been found. The relative signs of the principal values may be obtained by recording the ENDOR spectrum at different field positions between a $g_\perp$- and a $g_\parallel$-peak. These spectra demonstrate that the three hfs components must have the same sign, thus $a_{iso}^{CH_3} \simeq 1.0$ MHz. A similar technique for sign determination has been applied to the CH-protons with $A_3^{CH} = 3.2$ MHz (in-plane) and $A_2^{CH} = 1.7$ MHz (out-of-plane). The sign of A_2^{CH} was established by DOUBLE ENDOR to be opposite to $A_2^{CH_3}$. The second hfs component in the complex plane could not be measured. It was found by van Willigen[289, 291] that complexation of the sixth coordination position has a strong effect on the proton hf parameters and causes the methyl protons to become nonequivalent. The solvent effect on the proton hf splittings is, however, not fully understood at present.

VO(II)-Doped Triglycine Sulfate

In VO(II)-doped TGS, a "ferroelectric" splitting due to an interaction of the electric moment of the VO(II) ion with the internal electric field is observed at temperatures below the transition point $T_c = 49$ °C. Two different EPR spectra $(VO(II))_1$ and $(VO(II))_2$ of the thermally stable VO(II) complex are detected in TGS[292]. Site $(VO(II))_1$, with the more intense EPR spectrum, has been studied with ENDOR by Böttcher et al.[60]. The hfs tensors of the NH_2- and CH_2-protons are listed in Table 11. For the assignment of the four NH_2- and the four CH_2-protons to their corresponding N and C atoms, the authors made use of the nuclear dipole-dipole splitting and of the electron-nuclear point-dipole formula (5.6), respectively. In contrast to Cu(II)-doped TGS (Sect. 6.1.3)[188], the VO(II) ion is shifted by about 0.4 Å from the center of the straight line $\overline{N(II)N(III)}$, so that the hfs tensors of corresponding CH_2-protons in glycine II and III become strongly nonequivalent and an assignment is no longer ambiquous. The deviation of the hfs tensors from axiality is much smaller for the amino protons in VO(II): TGS than in Cu(II): TGS. This is due to the fact that in the former case the spin density on the nitrogen ligands is smaller than in Cu(II): TGS. Thus, the point-dipole formula is a suitable approximation for VO(II): TGS, in which the unpaired electron spin is located mainly on the VO(II) ion. The proton ENDOR data published by Böttcher et al.[60] confirm the earlier findings[292] that in VO(II)-doped TGS the vanadyl ion occupies an interstitial lattice site forming a $VO(gly)_2$ complex with triclinic symmetry C_1.

6.7 CrO(III) Compounds (d^1)

The proton ENDOR study of the chromyl ethyleneglycolate anion in ethanol reported by Möhl et al.[293] presents the first successful adaptation of the ENDOR technique to a transition metal complex in *liquid solutions*. The aim of this work was to characterize the ENDOR relaxation behavior and to find optimum conditions for ENDOR detection. Two proton ENDOR lines with a hf splitting of $a_{iso}^{H} = 1.74$ MHz were observed. This is in agreement with a previous EPR study[294] which had shown that all eight protons are equivalent. The optimum microwave and rf fields are both proportional to $(\mathrm{Tr}(\mathbf{g} - g_e\mathbf{I})^2 \mathrm{Tr} A_d^{H\,2})^{1/2}$, where A_d^H denotes the dipolar part of the proton hfs tensor. For the chromyl ethyleneglycolate anion these two values have been calculated to $B_{Mw}^{opt} \simeq 8 \cdot 10^{-6}$T and $B_{rf}^{opt} \simeq 2.7$ mT. According to Möhl et al.[293], successful proton

ENDOR on transition metal complexes in solutions is only attainable if no other nuclei possessing a much larger hf anisotropy than the protons are present. Moreover, the deviation of the $\mathbf{g}$ tensor principal values from g_e should be small, so that $\mathrm{Tr}(\mathbf{g} - g_e\mathbf{I})^2 < 3 \cdot 10^{-3}$. Solvent and temperature, however, appear to have minor influence on optimum ENDOR detection conditions.

7 Concluding Remarks

In this monograph it has been demonstrated that ENDOR is a very powerful tool to study transition metal ions in organic, inorganic and bioinorganic single crystals, polycrystalline samples and frozen solutions. Due to the high resolution of this double resonance technique, many problems in magnetic resonance, which cannot be solved with ordinary EPR, become accessible.

The various special ENDOR techniques summarized in Sect. 4 widen the field of applications considerably. They allow investigations either of complex, oriented spin systems, or of paramagnetic centers in randomly oriented large molecules. The ENDOR techniques are particularly useful to study biochemical systems, which are often characterized by very poorly resolved powder EPR spectra.

In the last few years, the number of ENDOR publications on systems of biological interest has remarkably increased, indicating that the advantages offered by this technique have been realized by many research groups. It should not be concealed, however, that in this type of compounds poor sensitivity may cause serious problems, so that large samples and (or) computer interfacing capabilities have usually to be available.

The ENDOR techniques, of course, are not confined to studies of transition metal complexes. A fast growing interest on electron nuclear double and multiple resonance experiments is also noticed in other fields of natural sciences, such as radical, radiation and polymer chemistry, solid state physics, biophysics and mineralogy.

Acknowledgement. I am indebted to Professor Hs. H. Günthard for his encouragement and steady interest in this work. Also, I wish to express my sincere thanks to my research associates Dr. M. Rudin, Mr. E. Jörin, Mr. R. Wolf and El. eng. J. Forrer who have collaborated with me on ENDOR problems with great enthusiasm. I am very grateful to Dr. G. Cirelli for careful reading of the manuscript and to Miss C. Rey for the drawing of the figures. Finally, I thank those who generously supplied me with information prior to publication.

Appendix A
(Abbreviations Used in This Paper)

1) Spectroscopy and Electronics

EPR	Electron Paramagnetic Resonance
ENDOR	Electron Nuclear Double Resonance
EI-EPR	ENDOR-induced EPR
CP-ENDOR	ENDOR with circularly polarized radio frequency fields
PM-ENDOR	Polarization modulated ENDOR
ESE	Electron Spin Echo
NQR	Nuclear Quadrupole Resonance
NMR	Nuclear Magnetic Resonance
MIMARS	Microwave Frequency-Magnetic Field Ratio Stabilization
PSD	Phase Sensitive Detector
VSWR	Voltage-Standing-Wave Ratio
APC	Automatic Phase Control
ALC	Automatic Level Control
CVC	Current-to-Voltage Converter
rf	radio frequency
MW	Microwave
cw	continuous wave
r.h.	right hand
l.h.	left hand
T (= 10^4 G)	Tesla
mT	Millitesla
iso	isotropic
av	average
opt	optimum
eff	effective
rot	rotating frame
hf	hyperfine
hfs	hyperfine splitting
NZ	Nuclear Zeeman
EZ	Electron Zeeman
EFG	Electric Field Gradient
SQT	Single Quantum Transition
DQT	Double Quantum Transition
MQT	Multiple Quantum Transition

2) Quantum Chemistry

MO	Molecular Orbital
TD	Townes-Dailey
INDO	Intermediate Neglect of Differential Overlap
EHT	Extended Hückel Theory
SCCC	Self Consistence of Charge and Configuration

3) Ligands and Host Compounds

Ox	oxychinolate
HOx	8-hydroxyquinoline
PHI	phthalimide
PHA	phthalic acid
sal	salicylaldoxime
msal	methyl-salicylaldoxime
acacen	N,N′-ethylene-bis(acetylacetonatiminate)
amben	N,N′-ethylene-bis (*o*-aminobenzylideniminate)
salen	N,N′-ethylene-bis(salicylideniminate)
3 MeOsalen	3-methoxy-salen
4 Mesalen	4-methyl-salen
5 MeOsalen	5-methoxy-salen
salim	salicylaldehyde-imine
et_2dtc	N,N′-diethyl-dithiocarbamate
acac	2,4-pentanedionate
dim	dimethylglyoxime
pic	picolinate
Me(pic)	2-methylpicolinate
qn	quinaldinate
TPP	tetraphenylporphyrin
gly	glycine
bipyam	di-2-pyridylamine
dien	diethylenetriamine
TGS	triglycine sulfate
Mb	myoglobin
Hb	hemoglobin
bz	benzene
cp	cyclopentadienyl
py	pyridine
THF	tetrahydrofurane

Appendix B
(Second Order ENDOR Frequencies)

In order to arrive at clearly structured expressions the following notation concerning division of matrices is adopted:

$$\mathbf{X} = \left[\begin{array}{ccc} X_{11} & X_{12} & X_{13} \\ X_{21} & X_{22} & X_{23} \\ \hline X_{31} & X_{32} & X_{33} \end{array}\right] = \left[\begin{array}{c|c} \mathbf{X}^{11} & \mathbf{X}^{13} \\ \hline \mathbf{X}^{31} & X_{33} \end{array}\right] = \left[\begin{array}{cc|c} X_{11} & X_{12} & X_{13} \\ X_{21} & X_{22} & X_{23} \\ X_{31} & X_{32} & X_{33} \end{array}\right] = (\mathbf{X}^{12}\ \mathbf{X}_{.3})\,. \quad \text{(B 1)}$$

Elements of the rotation matrix $\mathbf{R}^{S}$:

$$\begin{aligned} \mathbf{R}^{S,11}_{1.} &= \frac{1}{g g_{12}} \tilde{\mathbf{R}}_{.3}\, g_{.3}\, \tilde{\mathbf{R}}_{.3}\, \mathbf{g}^{12} \\ \mathbf{R}^{S,11}_{2.} &= \frac{1}{g_{12}} \tilde{\mathbf{R}}_{.3}\, \mathbf{g}^{12} \begin{pmatrix} 0 & 1 \\ -1 & 0 \end{pmatrix} \\ \mathbf{R}^{S,13} &= -\frac{g_{12}}{g} \begin{pmatrix} 1 \\ 0 \end{pmatrix} \\ \mathbf{R}^{S}_{3.} &= \frac{1}{g} \tilde{\mathbf{R}}_{.3}\, \mathbf{g} \end{aligned} \quad \text{(B 2)}$$

with $g_{12} = |(\tilde{\mathbf{R}}_{.3}\mathbf{g}^{12}\tilde{\mathbf{g}}^{12}\mathbf{R}_{.3})^{1/2}|$.

$\mathbf{R}^{S,11}_{i.}$ and $\mathbf{R}^{S}_{3.}$ denote the ith row of the 2×2 matrix $\mathbf{R}^{S,11}$ and the third row of the matrix $\mathbf{R}^{S}$, respectively.

Elements of the rotation matrix $\mathbf{R}^{I}(m_S)$:

$$\begin{aligned} \mathbf{R}^{I,11}_{1.}(m_S) &= \frac{1}{c_{12}(m_S)c(m_S)} \tilde{\mathbf{R}}_{.3}\, \mathbf{C}_{.3}(m_S)\tilde{\mathbf{R}}_{.3}\, \mathbf{C}^{12}(m_S) \\ \mathbf{R}^{I,11}_{2.}(m_S) &= \frac{1}{c_{12}(m_S)} \tilde{\mathbf{R}}_{.3}\, \mathbf{C}^{12}(m_S) \begin{pmatrix} 0 & 1 \\ -1 & 0 \end{pmatrix} \\ \mathbf{R}^{I,13}(m_S) &= -\frac{c_{12}(m_S)}{c(m_S)} \begin{pmatrix} 1 \\ 0 \end{pmatrix} \end{aligned} \quad \text{(B 3)}$$

$$\mathbf{R}^{\mathrm{I}}_{3.}(m_S) = \frac{1}{c(m_S)}\,\tilde{\mathbf{R}}_{.3}\,\mathbf{C}(m_S)$$

with $c_{12}(m_S) = |(\tilde{\mathbf{R}}_{.3}\,\mathbf{C}^{12}(m_S)\,\tilde{\mathbf{C}}^{12}(m_S)\,\mathbf{R}_{.3})^{1/2}|$.

Second order ENDOR transition frequencies ($m_S = 1/2$):

a) direct product base:
One I = 1 ligand nucleus in the presence of a second nucleus (K = 1)

$$\begin{aligned}
\Delta E(1 \leftrightarrow 0):\quad & c + \frac{3}{2}\,\overline{Q}_{33} + \beta(-a_1 + 2a_2 + 2a_3 - 4b_3) + q' \\
& c + \frac{3}{2}\,\overline{Q}_{33} + \beta(-a_1 + 2a_2 + 2a_3) + q' \\
& c + \frac{3}{2}\,\overline{Q}_{33} + \beta(-a_1 + 2a_2 + 2a_3 + 4b_3) + q' \\
\Delta E(0 \leftrightarrow -1):\quad & c - \frac{3}{2}\,\overline{Q}_{33} + \beta(+a_1 + 2a_2 - 2a_3 - 4b_3) + q' \\
& c - \frac{3}{2}\,\overline{Q}_{33} + \beta(+a_1 + 2a_2 - 2a_3) + q' \\
& c - \frac{3}{2}\,\overline{Q}_{33} + \beta(+a_1 + 2a_2 - 2a_3 + 4b_3) + q' \,.
\end{aligned} \tag{B 4}$$

Equation (B 4) may easily be extended to an arbitrary number of nuclei. In this case, cross-terms of the type $4\beta\,b_3$ between each pair of nuclei occur.

b) Coupled spin base:
Two magnetically equivalent I = 1 nuclei (F = 0, 1, 2)

F	$\Delta E(F, \overline{F}, \overline{m}_F \leftrightarrow m_F)$	rel. Int.
1	$\Delta E(1 \leftrightarrow 0) = c - \frac{3}{2}\,\overline{Q}_{33} + \beta(-a_1 + 2a_2 + 2a_3) + q'$	1
1	$\Delta E(0 \leftrightarrow -1) = c + \frac{3}{2}\,\overline{Q}_{33} + \beta(+a_1 + 2a_2 - 2a_3) + q'$	1
2	$\Delta E(2 \leftrightarrow 1) = c + \frac{3}{2}\,\overline{Q}_{33} + \beta(-3a_1 + 2a_2 + 6a_3) + q'$	2
2	$\Delta E(1 \leftrightarrow 0) = c - \frac{3}{2}\,\overline{Q}_{33} + \beta(+3a_1 + 2a_2 + 2a_3) + q'$	1

(B 5)

F	$\Delta E(F, \overline{F}, \overline{m}_F \leftrightarrow m_F)$	rel. Int.
2	$\Delta E(0 \leftrightarrow -1) = c + \frac{3}{2}\overline{Q}_{33} + \beta(-3a_1 + 2a_2 - 2a_3) + q'$	1
2	$\Delta E(-1 \leftrightarrow -2) = c - \frac{3}{2}\overline{Q}_{33} + \beta(+3a_1 + 2a_2 - 6a_3) + q'$	2
$0 \leftrightarrow 2$	$\Delta E(1 \leftrightarrow 0) = c + \frac{3}{2}\overline{Q}_{33} + \beta(+a_1 + 2a_2 + 2a_3) + q'$	2
$0 \leftrightarrow 2$	$\Delta E(0 \leftrightarrow -1) = c - \frac{3}{2}\overline{Q}_{33} + \beta(-a_1 + 2a_2 - 2a_3) + q'$	2

c) Symmetrized spin base:
Two geometrically equivalent I = 1 nuclei in symmetry planes

Symmetry Species A

$$
\begin{aligned}
& c + \frac{3}{2}\overline{Q}_{33} + \beta(-a_1 + 2a_2 + 2a_3 - 4b_3 - 2b_1) + q' \\
& c + \frac{3}{2}\overline{Q}_{33} + \beta(-a_1 + 2a_2 + 2a_3 + 2b_1) + q' \\
& c + \frac{3}{2}\overline{Q}_{33} + \beta(-a_1 + 2a_2 + 2a_3 + 4b_3 - 2b_1) + q' \\
& c - \frac{3}{2}\overline{Q}_{33} + \beta(+a_1 + 2a_2 - 2a_3 - 4b_3 + 2b_1) + q' \\
& c - \frac{3}{2}\overline{Q}_{33} + \beta(+a_1 + 2a_2 - 2a_3 - 2b_1) + q' \\
& c - \frac{3}{2}\overline{Q}_{33} + \beta(+a_1 + 2a_2 - 2a_3 + 4b_3 + 2b_1) + q'
\end{aligned}
\tag{B 6}
$$

Symmetry Species B

$$
\begin{aligned}
& c + \frac{3}{2}\overline{Q}_{33} + \beta(-a_1 + 2a_2 + 2a_3 - 4b_3 + 2b_1) + q' \\
& c - \frac{3}{2}\overline{Q}_{33} + \beta(+a_1 + 2a_2 - 2a_3 + 4b_3 - 2b_1) + q'
\end{aligned}
$$

Appendix C
(Relations Between Nuclear Quadrupole Coupling Constants in Different Expressions of $\mathcal{H}_Q$ (Sect. 5.2))

The pertubation approach given in Sect. 3 yields the principal values and the direction cosines of the quadrupole tensor **Q** which writes in its principal axes frame:

$$\mathbf{Q} = \begin{pmatrix} Q_1 & & \\ & Q_2 & \\ & & Q_3 \end{pmatrix} \quad \text{with } (Q_1 + Q_2 + Q_3) = 0 \text{ and}$$

$$\eta = (Q_1 - Q_2)/Q_3, \quad |Q_3| > |Q_2| > |Q_1|$$

The coupling constants in Eqs. (5.16 a), (5.16 b) and (5.16 c) are related by:

Eqs. (5.16 a) $\leftrightarrow$ (5.16 b):

$$Q_1 = \frac{1}{3} Q' (\eta - 1)$$

$$Q_2 = -\frac{1}{3} Q' (\eta + 1)$$

$$Q_3 = \frac{2}{3} Q'$$

Eqs. (5.16 b) $\leftrightarrow$ (5.16 c):

$$Q' = \frac{3}{4} \frac{e^2qQ}{I(2I - 1)h}$$

$$Q'' = \frac{e^2qQ\eta}{8I(2I - 1)h}$$

For frequently occurring nuclear spin quantum numbers, the relations are given by:

I = 1 (e.g. ^{2}D, ^{14}N):

$$Q' = \frac{3e^2qQ}{4h}, \quad Q'' = \frac{e^2qQ\eta}{8h}, \quad \frac{e^2qQ}{h} = 2Q_3 \text{ (Q_3: largest principal value)}$$

For this nuclear quantum number the frequencies

$$\nu_\pm = \frac{3e^2qQ}{4h}\left(1 \pm \frac{\eta}{3}\right) \text{ measured in NQR are related to the principal values of } \mathbf{Q} \text{ by}$$

$$\nu_+ = Q_3 - Q_2$$

$$\nu_- = Q_3 - Q_1$$

I = 3/2 (e.g. $^{35,37}Cl$, $^{63,65}Cu$):

$$Q' = \frac{e^2qQ}{4\,h}, \quad Q'' = \frac{e^2qQ\eta}{24\,h}, \quad \frac{e^2qQ}{h} = 6\,Q_3$$

I = 7/2 (e.g. ^{59}Co, ^{51}V):

$$Q' = \frac{3\,e^2qQ}{84\,h}, \quad Q'' = \frac{e^2qQ\eta}{168\,h}, \quad \frac{e^2qQ}{h} = 42\,Q_3$$

References

1. Owen, J., Stevens, K. W. H.: Nature *171*, 836 (1953)
2. Abragam, A., Bleaney, B.: Electron Paramagnetic Resonance, of Transition Ions. Oxford: Clarendon Press (1970)
3. Feher, G.: Phys. Rev. *103*, 834 (1956)
4. Kevan, L., Kispert, L. D.: Electron Spin Double Resonance Spectroscopy. New York: John Wiley and Sons (1976)
5. Dorio, M. M., Freed, J. H., eds.: Multiple Electron Resonance Spectroscopy. New York: Plenum Press (1979)
6. Box, H. C.: Radiation Effects: ESR and ENDOR Analysis. New York: Academic Press (1977)
7. Hyde, J. S.: in: Magnetic Resonance in Biological Systems. Ehrenberg, A., Malmström, B. G., Vänngard, T., eds. London: Pergamon (1967) p. 63
8. Hyde, J. S.: Ann. Rev. Phys. Chem. *25*, 407 (1974)
9. Baker, J. M., Davies, E. R., Reddy, T. R.: Contemp. Phys. *13*, 45 (1972)
10. Kwiram, A. L.: Ann. Rev. Phys. Chem. *22*, 133 (1971)
11. Norris, J. R., Scheer, H., Katz, J. J.: in The Porphyrins, Vol. 4. Dolphin, D., ed. New York: Academic Press (1979) p. 159
12. Rudin, M., Schweiger, A., Günthard, Hs. H.: Mol. Phys., to be published
13. Deal, R. M., Ingram, D. J. E., Srinivasan, R.: Electric Magnetic Resonance and Solid Dielectrics, Proc. Colloq. AMPERE, 12th, p. 239 (1963)
14. Eisenberger, P., Pershan, P. S.: J. Chem. Phys. *47*, 3327 (1967)
15. Rist, G. H., Hyde, J. S.: J. Chem. Phys. *50*, 4532 (1969)
16. Rist, G. H., Hyde, J. S.: J. Chem. Phys. *52*, 4633 (1970)
17. Rist, G. H., Hyde, J. S., Vänngard, T.: Proc. Nat. Acad. Sci. USA *67*, 79 (1970)
18. Leniart, D. S.: in: Multiple Electron Resonance Spectroscopy. Dorio, M. M., Freed, J. H., eds. New York: Plenum Press (1979)
19. Nelson, W. H.: J. Mag. Res. *37*, 205 (1980)
20. Cook, R. J.: J. Sci. Instr. *43*, 548 (1966)
21. Becker, D., Kwiram, A. L.: Chem. Phys. Letters *39*, 180 (1976)
22. Yagi, H., Inoue, M., Tatsukawa, T., Yamamoto, T.: Jap. J. App. Phys. *9*, 1386 (1970)
23. Hyde, J. S.: J. Chem. Phys. *43*, 1806 (1965)
24. Schmalbein, D., Witte, A., Roder, R., Laukien, G.: Rev. Sci. Instr. *43*, 1664 (1972)
25. Dinse, K. P., Möbius, K., Biehl, R.: Z. Naturforsch. *28A*, 1069 (1973)
26. Miyagawa, I., Davidson, R. B., Helms, H. A., Wilkinson, B. A.: J. Mag. Res. *10*, 156 (1973)
27. Gruber, K., Forrer, J., Schweiger, A., Günthard, Hs. H.: J. Phys. *E7*, 569 (1974)
28. Hyde, J. S., Maki, A. H.: J. Chem. Phys. *40*, 3117 (1964)
29. Yamamoto, T., Kono, M., Sato, K., Miyamae, T., Mukai, K., Ishizu, K.: JEOL News *10a*, 6 (1972)
30. Varian Instrument Division Bull., E-700 High Power ENDOR System (1971)
31. Berchten, N.: unpublished
32. Hyde, J. S., Astlind, T., Eriksson, L. E. G., Ehrenberg, A.: Rev. Sci. Instr. *41*, 1598 (1970)
33. Varian field/frequency lock E-272B
34. Andrist, M., Loth, K., Günthard, Hs. H.: J. Phys. *E9*, 947 (1976)
35. Niklas, J. R., Spaeth, J. M.: Phys. Stat. Sol. *B101*, 221 (1980)
36. Kirste, B., van Willigen, H., Kurreck, H., Möbius, K., Plato, M., Biehl, R.: J. Am. Chem. Soc. *100*, 7505 (1978)
37. Schweiger, A., Rudin, M., Forrer, J.: Chem. Phys. Letters *80*, 376 (1981)

38. Gächter, B.: Thesis No. 4909, ETH Zürich (1972)
39. Oppliger, H. R., Schmid, P., Günthard, Hs. H.: J. Sci. Instr. *42*, 865 (1965)
40. Schweiger, A., Rudin, M., Günthard, Hs. H.: Mol. Phys. *41*, 63 (1980)
41. Rudin, M., Schweiger, A., Günthard, Hs. H.: Chem. Phys. Letters, to be published
42. Forrer, J., Schweiger, A., Günthard, Hs. H.: J. Phys. *E 10*, 470 (1977)
43. Forrer, J., Schweiger, A., Berchten, N., Günthard, Hs. H.: J. Phys. *E 14*, 565 (1981)
44. Schweiger, A., Günthard, Hs. H.: Chem. Phys. Letters, to be published
45. Schweiger, A., Forrer, J., Günthard, Hs. H.: J. Mag. Res., to be published
46. Abragam, A., Pryce, M. H. L.: Proc. Roy. Soc. *A 205*, 135 (1951)
47. Weil, J. A., Anderson, J. H.: J. Chem. Phys. *35*, 1410 (1961)
48. Weil, J. A.: J. Mag. Res. *18*, 113 (1975)
49. Iwasaki, M.: J. Mag. Res. *16*, 417 (1974)
50. Thuomas, K., Lund, A.: J. Mag. Res. *18*, 12 (1975)
51. Rockenbauer, A., Simon, P.: J. Mag. Res. *11*, 217 (1973)
52. Rockenbauer, A., Simon, P.: Mol. Phys. *28*, 1113 (1974)
53. Zevin, V. Y., Ishchenko, S. S., Ruban, M. A.: Soviet Phys. JETP *28*, 1116 (1969)
54. Hagston, W. E., Holmes, B. J.: J. Phys. *B 13*, 3505 (1980)
55. Schweiger, A., Graf, F., Rist, G., Günthard, Hs. H.: Chem. Phys. *17*, 155 (1976)
56. Atherton, N. M., Shackleton, J. F.: Mol. Phys. *39*, 1471 (1980)
57. Snaathorst, D., Keijzers, C. P., Klaassen, A. A. K., de Boer, E., Chacko, V. P., Gomperts, R.: Mol. Phys. *40*, 585 (1980)
58. Fujimoto, M., McDowell, C. A., Takui, T.: J. Chem. Phys. *70*, 3694 (1979)
59. Rudin, M., Jörin, E., Schweiger, A., Günthard, Hs. H.: Chem. Phys. Letters *67*, 374 (1979)
60. Böttcher, R., Heinhold, D., Windsch, W.: Chem. Phys. Letters *65*, 452 (1979)
61. Rudin, M., Schweiger, A., Berchten, N., Günthard, Hs. H.: Mol. Phys. *41*, 1317 (1980)
62. Schweiger, A., Günthard, Hs. H.: Chem. Phys. *32*, 35 (1978)
63. Calvo, R., Oseroff, S. B., Abache, H. C.: J. Chem. Phys. *72*, 760 (1980)
64. Teslenko, V. V., Gromovoi, Y. S., Krivenko, V. G.: Mol. Phys. *30*, 425 (1975)
65. Schweiger, A., Rist, G., Günthard, Hs. H.: Chem. Phys. Letters *31*, 48 (1975)
66. Brown, T. G., Hoffman, B. M.: Mol. Phys. *39*, 1073 (1980)
67. Schweiger, A., Rudin, M., Günthard, Hs. H.: Mol. Phys. *37*, 1573 (1979)
68. Rudin, M., Schweiger, A.: unpublished results
69. Abragam, A.: The Principle of Nuclear Magnetism. Oxford: Clarendon Press (1961)
70. Mulks, C. F., Scholes, C. P., Dickinson, L. C., Lapidot, A.: J. Am. Chem. Soc. *101*, 1645 (1979), supplementary material
71. Mulks, C. F., van Willigen, H.: J. Phys. Chem. *85*, 1220 (1981)
72. Coope, J. A. R., Dalal, N. S., McDowell, C. A., Srinivasan, R.: Mol. Phys. *24*, 403 (1972)
73. Dinse, K. P., Möbius, K., Plato, M., Biehl, R., Haustein, H.: Chem. Phys. Letters *14*, 196 (1972)
74. DuVarney, R. C., Spaeth, J. M.: Solid State Commun. *32*, 1237 (1979)
75. Rist, G. H., Hyde, J. S.: J. Chem. Phys. *49*, 2449 (1968)
76. Rist, G. H.: Habilitationsschrift, ETHZ
77. Dalton, L. R., Kwiram, A. L.: J. Chem. Phys. *57*, 1132 (1972)
78. Schweiger, A., Wolf, R., Günthard, Hs. H., Ammeter, J. H., Deiss, E.: Chem. Phys. Letters *71*, 117 (1980)
79. van Camp, H. L., Scholes, C. P., Mulks, C. F., Caughey, W. S.: J. Am. Chem. Soc. *99*, 8283 (1977)
80. Jörin, E., Rudin, M., Schweiger, A., Günthard, Hs. H.: Chem. Phys. Letters *69*, 193 (1980)
81. Ovchinnikov, I. V., Konstantinov, V. N.: J. Mag. Res. *32*, 179 (1978)
82. Schwerdtfeger, C. F., Diehl, P.: Mol. Phys. *17*, 417 (1969)
83. Fackler, J. P., Levy, J. D., Smith, J. A.: J. Am. Chem. Soc. *94*, 2436 (1972)
84. Rudin, M., Schweiger, A., Günthard, Hs. H.: to be published
85. Allendoerfer, R. D., Papez, R. J.: J. Am. Chem. Soc. *92*, 6971 (1970)
86. Biehl, R., Lubitz, W., Möbius, K., Plato, M.: J. Chem. Phys. *66*, 2074 (1977)
87. Robinson, B. H., Dalton, L. A., Beth, A. H., Dalton, L. R.: Chem. Phys. *18*, 321 (1976)
88. Cook, R. J., Whiffen, D. H.: Proc. Phys. Soc. *84*, 845 (1964)
89. Freed, J. H.: J. Chem. Phys. *50*, 2271 (1969)

90. Dinse, K. P., Biehl, R., Möbius, K.: J. Chem. Phys. *61*, 4335 (1974)
91. Biehl, R., Plato, M., Möbius, K.: J. Chem. Phys. *63*, 3515 (1975)
92. Biehl, R., Möbius, K., O'Connor, S. E., Walter, R. I., Zimmermann, H.: J. Phys. Chem. *83*, 3449 (1979)
93. Möbius, K., Biehl, R.: In: Multiple Electron Resonance Spectroscopy, Dorio, M. M., Freed, J. H., eds., New York: Plenum Press (1979)
94. Baker, J. M., Blake, W. B. J.: Phys. Letters *31 A*, 61 (1970)
95. Baker, J. M., Blake, W. B. J.: J. Phys. C.: Solid St. Phys. *6*, 3501 (1973)
96. Anderson, W. A., Nelson, F. A.: J. Chem. Phys. *39*, 183 (1963)
97. Haeberlen, U.: Advances in Magnetic Resonance, Waugh, J. S., ed. New York: Academic Press (1976)
98. Anderson, W. A., Freeman, R.: J. Chem. Phys. *37*, 85 (1962)
99. Dalal, N. S., Manoogian, A.: Phys. Rev. Letters *39*, 1573 (1977)
100. Manoogian, A., Chan, B. W.: Can. J. Phys. *58*, 687 (1980)
101. Balmer, P., Blum, H., Forster, M., Schweiger, A., Günthard, Hs. H.: J. Phys. C: Solid St. phys. *13*, 517 (1980)
102. Freed, J. H., Leniart, D. S., Hyde, J. S.: J. Chem. Phys. *47*, 2762 (1967)
103. Dinse, K. P., Biehl, R., Möbius, K., Plato, M.: J. Mag. Res. *6*, 444 (1972)
104. Schweiger, A., Günthard, Hs. H.: Mol. Phys. *42*, 283 (1981)
105. Power, C. P.: Ph. D. Thesis, University College, Cork (1975)
106. Schweiger, A., Günthard, Hs. H., to be published
107. Blume, R. J.: Phys. Rev. *109*, 1867 (1958)
108. Mims, W. B.: "Electron spin echoes" in: Electron Paramagnetic Resonance, ed. Geschwind, S., New York: Plenum Press (1972), chapt. 4
109. Norris, J. R., Thurnauer, M. C., Bowman, M. K.: Adv. Biol. Med. Phys.
110. Kevan, L., Schwartz, R. N.: Time Domain Electron Spin Resonance, New York: Wiley (1979)
111. Rowan, L. G., Hahn, E. L., Mims, W. B.: Phys. Rev. *A 137*, 61 (1965)
112. Mims, W. B.: Phys. Rev. *B 5*, 2409 (1972)
113. Merks, R. P. J., de Beer, R.: J. Magn. Res. *37*, 305 (1980)
114. de Beer, R.: unpublished results
115. Merks, R. P. J., de Beer, R.: J. Phys. Chem. *83*, 3319 (1979)
116. van Ormondt, D., Nederveen, K.: Chem. Phys. Letters *82*, 443 (1981)
117. Peisach, J.: Dev. Biochem. 285 (1978)
118. Mims, W. B.: Proc. Roy. Soc. *A 283*, 452 (1965)
119. de Beer, R., Barkhuijsen, H., de Wild, E. L., Merks, R. P. J.: Bull. Mag. Res. *2*, 420 (1981)
120. Kuska, H. A., Rogers, M. T.: Spect. Inorg. Chem. *2*, 175 (1971)
121. Owen, J., Thornley, J. H. M.: Rep. Prog. Phys. *29*, 675 (1966)
122. McGarvey, B. R.: in: Transition Metal Chemistry, Vol. 3, Carlin, R. L., ed. New York: Marcel Dekker (1966) p. 90
123. Electron Spin Resonance of Metal Complexes; Yen, T. F., ed. New York: Plenum Press (1969)
124. Freeman, A. J., Watson, R. E.: in: Magnetism, Vol. 2 A, Rado, G., Suhl, H., eds. New York: Academic Press (1965) p. 167
125. Marshall, W.: in: Paramagnetic Resonance, Low, W., ed. New York: Academic Press (1963) p. 347
126. Keijzers, C. P., de Boer, E.: J. Chem. Phys. *57*, 1277 (1972)
127. Keijzers, C. P., Snaathorst, D.: Chem. Phys. Letters *69*, 348 (1980)
128. Han, P. S., Das, T. P., Rettig, M. F.: Theoret. Chim. Acta. *16*, 1 (1970)
129. Haberditzl, W.: Quantenchemie, Vol. 4, Heidelberg: Dr. Alfred Hüthig Verlag (1979)
130. de Beer, R., de Boer, W., Van't Hof, C. A., van Ormondt, D.: Acta Cryst. *B 29*, 1473 (1973)
131. van Ormondt, D., de Beer, R., Brouha, M., de Groot, F.: Z. Naturforsch. *24 a*, 1746 (1969)
132. Atherton, N. M., Horsewill, A. J.: J. Chem. Soc. Faraday II *76*, 660 (1980)
133. Wertz, J. E., Bolton, J. R.: Electron Spin Resonance, New York: McGraw-Hill (1972)
134. McConnell, H. M.: Proc. Natl. Acad. Sci., USA *44*, 766 (1958)
135. Kneubühl, F. K.: Phys. Kondens. Mater *1*, 410 (1963)
136. Kneubühl, F. K.: Phys. Kondens. Mater *4*, 50 (1965)

137. Merks, R. P. J., van der Toorn, J. D., Toledo, B. J., de Beer, R.: in: Magnetic Resonance and Related Phenomena: Proc. 18th Ampere Congress, Nottingham, Allen, P. S., Andrew, E. R., Bates, C. A., eds. Amsterdam: Elsevier (1974) p. 95
138. Calvo, R.: J. Mag. Res. *26*, 445 (1977)
139. Rockenbauer, A., Simon, P.: J. Mag. Res. *22*, 243 (1976)
140. Golding, R. M., Stubbs, L. C.: Proc. Roy. Soc. Lond. *A 354*, 223 (1977)
141. Belford, R. L., Pilbrow, J. R.: J. Mag. Res. *11*, 381 (1973)
142. Pilbrow, J. R., Lowrey, M. R.: Rep. Prog. Phys.: *43*, 434 (1980)
143. McConnell, H. M., Robertson, R. E.: J. Chem. Phys. *29*, 1361 (1958)
144. Hutchison, C. A., McKay, D. B.: J. Chem. Phys. *66*, 3311 (1977)
145. Maki, A. H., McGarvey, B. R.: J. Chem. Phys. *29*, 31 (1958)
146. Ammeter, J.: Ph. D. Thesis ETHZ Nr. 4338 (1969)
147. Lorenz, A.: Ph. D. Thesis ETHZ Nr. 5400 (1974)
148. Gerloch, M., Miller, J. R.: Progr. Inorg. Chem. *10*, 1 (1968)
149. Slichter, C. P.: Principles of Magnetic Resonance, 2nd ed. New York: Springer (1978)
150. Lucken, E. A. C.: Nuclear Quadrupole Coupling Constants, New York: Academic Press (1969)
151. Townes, C. H., Dailey, B. P.: J. Chem. Phys. *17*, 782 (1949)
152. Kochanski, E., Lehn, J. M., Levy, B.: Theoret. Chim. Acta. *22*, 111 (1971)
153. van Bronswyk, W.: Structure and Bonding *7*, 87 (1970)
154. Brown, T. L.: Acc. Chem. Res. *7*, 408 (1974)
155. Brown, T. L.: J. Mol. Struct. *58*, 293 (1980)
156. Brill, T. B.: Advances in Nuclear Quadrupole Resonance, Vol. 3, Smith, J. A. S., ed. London: Heyden (1978) p. 131
157. White, L. K., Belford, R. L.: J. Am. Chem. Soc. *98*, 4428 (1976)
158. Kita, S., Hashimoto, M., Iwaizumi, M.: to be published
159. Chacko, V. P., Manoharan, P. T.: J. Mag. Res. *22*, 7 (1976)
160. Han, P. S., Rettig, M. F., Ikenberry, D., Das, T. P.: Theoret. Chim. Acta (Berlin) *22*, 261 (1971)
161. Hsieh, Y. N., Ireland, P. S., Brown, T. L.: J. Mag. Res. *21*, 445 (1976)
162. Goodman, B. A., Raynor, J. B.: Advan. Inorg. Chem. Rad. Chem. *13*, 135 (1970)
163. Kuska, H. A., Rogers, M. T.: Radical Ions; Kaiser, E. T., Kevan., L., eds. New York: Interscience (1968) p. 579
164. Jarski, M. A., Lingafelter, E. C.: Acta Crystallogr. *17*, 1109 (1964)
165. Nakamoto, K., Margoshes, M., Rundle, E.: J. Am. Chem. Soc. *77*, 6480 (1955)
166. Pimentel, G. C., McClellan, A. L.: Ann. Rev. Phys. Chem. *22*, 347 (1971)
167. Schweiger, A.: Ph. D. Thesis ETHZ Nr. 5748 (1976)
168. Moores, B. W., Belford, R. L.: in: Electron Spin Resonance of Metal Complexes, Yen, T. F., ed. New York: Plenum Press, (1969) p. 13
169. Murgich, J.: J. Chem. Phys. *74*, 3618 (1981)
170. Sauer, E. G., Bray, P. J.: J. Chem. Phys. *58*, 21 (1973)
171. Brown, T. G., Petersen, J. L., Lozos, G. P., Anderson, J. R., Hoffman, B. M.: Inorg. Chem. *16*, 1563 (1977)
172. Case, D. A., Karplus, M.: J. Am. Chem. Soc. *99*, 6182 (1977)
173. Kita, S., Hashimoto, M., Iwaizumi, M.: Inorg. Chem. *18*, 3432 (1979)
174. Kita, S., Iwaizumi, M.: Chem. Letters 67 (1979)
175. Maki, A. H., McGarvey, B. R.: J. Chem. Phys. *29*, 35 (1958)
176. Malatesta, V., McGarvey, B. R.: Can. J. Chem. *53*, 3791 (1975)
177. Iwaizumi, M., et al.: to be published
178. Kirmse, R., Stach, J., Böttcher, R.: Chem. Phys. Letters *75*, 565 (1980)
179. Bonamico, M., Dessy, G., Mugnoli, A., Vaciago, A., Zambonelli, L.: Acta Cryst. *19*, 886 (1965)
180. Bonamico, M., Mazzone, G., Vaciago, A., Zambonelli, L.: Acta Cryst. *19*, 898 (1965)
181. Windsch, W., Welter, M.: Z. Naturforsch. *22*, 1 (1967)
182. Fujimoto, M., Janecka, J.: J. Chem. Phys. *55*, 1152 (1971)
183. Fujimoto, M., Wylie, L. A., Saito, S.: J. Chem. Phys. *58*, 1273 (1973)
184. Takeda, K., Arata, Y., Fujiwara, S.: J. Chem. Phys. *53*, 854 (1970)

185. Simpson, H. J., Marsh, R. E.: Acta Cryst. *20*, 550 (1966)
186. Murgich, J., Calvo, R., Oseroff, S. B.: J. Mol. Struct. *68*, 203 (1980)
187. Windsch, W., Welter, M., Driesel, W.: Ferroelectrics *8*, 551 (1974)
188. Böttcher, R., Heinhold, D., Windsch, W.: Chem. Phys. Letters *49*, 148 (1977)
189. Böttcher, R., Heinhold, D., Wartewig, S., Windsch, W.: J. Mol. Struct. *46*, 363 (1978)
190. Hoshino, S., Okaya, Y., Pepinsky, R.: Phys. Rev. *115*, 323 (1959)
191. Windsch, W.: privat communication
192. Stach, J., Kirmse, R., Wartewig, S., Böttcher, R.: Z. Chem. *19*, 205 (1979)
193. Malmström, B., Mosbach, R., Vänngard, T.: Nature *183*, 321 (1959)
194. Vänngard, T.: in: Biological Applications of EPR; Swartz, H. M., Bolton, J. R., Borg, D. C., eds. New York: Wiley Interscience (1972)
195. Boas, J. F., Pilbrow, J. R., Smith, T. D.: in: Biological Magnetic Resonance I; Berliner, L. J., Reuben, J., eds. New York: Plenum Press (1980) chapt. 7
196. Rotilio, G.: in: ESR and NMR in Biological and Related Systems, eds.: Bertini, I., Drago, R. S., eds. Dordrecht: D. Reidel Publ. Comp. (1980)
197. Fee, J. A.: Structure and Bonding *23*, 1 (1975)
198. Brill, A. S.: Transition Metals in Biochemistry, Berlin: Springer (1977) p. 40
199. Roberts, J. E., Brown, T. G., Hoffman, B. M., Peisach, J.: J. Am. Chem. Soc. *102*, 825 (1980)
200. Hoffman, B. M., Roberts, J. E., Swanson, M., Speck, S. H., Margoliash, E.: Proc. Natl. Acad. Sci. USA *77*, 1452 (1980)
201. van Camp, H. L., Wei, Y. H., Scholes, C. P., King, T. E.: Biochim. Biophys. Acta *537*, 238 (1978)
202. Malmström, B. G., Reinhammer, B., Vänngard, T.: Biochim. Biophys. Acta *205*, 48 (1970)
203. Mims, W. B., Peisach, J.: Biochem. *15*, 3863 (1976)
204. Colman, P. M., Freeman, H. C., Guss, J. M., Murata, M., Norris, V. A., Ramshaw, J. A. M., Venkatappa, M. P.: Nature (London) *272*, 319 (1978)
205. Adman, E. T., Stenkamp, R. E., Sieker, L. C., Jensen, L. H.: J. Mol. Biol. *123*, 35 (1978)
206. McMillin, D. R.: Bioinorg. Chem. *8*, 179 (1978)
207. Caughey, W. S., Wallace, W. J., Volpe, J. A., Yoshikawa, S.: in: The Enzymes, Vol. 13, Boyer, P. D., ed. New York: Academic (1976) p. 299
208. Peisach, J., Blumberg, W. E.: Arch. Biochem. Biophys. *165*, 691 (1974)
209. Beinert, H., Griffiths, D. E., Wharton, D. C., Sands, R. H.: J. Biol. Chem. *237*, 2337 (1962)
210. Froncisz, W., Scholes, C. P., Hyde, J. S., Wei, Y. H., King, T. E., Shaw, R. W., Beinert, H.: J. Bio. Chem. *254*, 7482 (1979)
211. Froncisz, W., Hyde, J. S.: J. Chem. Phys. *73*, 3123 (1980)
212. Greenaway, F. T., Chan, S. H. P., Vincow, G.: Biochim. Biophys. Acta *490*, 62 (1977)
213. Chan, S. I., Bocian, D. F., Brudvig, G. W., Norse, R. H., Stevens, T. H.: in: Cytochrome Oxidase; King, T. E., Orii, Y., Change, B., Okunuki, K., eds. Amsterdam: Elsevier (1979) p. 177
214. Blumberg, W. E., Peisach, J.: in: Cytochrome Oxidase; King, T. E., Orii, Y., Change, B., Okunuki, K., eds. Amsterdam: Elsevier (1979) p. 153
215. Atherton, N. M., Horsewill, A. J.: Mol. Phys. *42*, 985 (1981)
216. van Niekerk, J. N., Schoening, F. R. L., Talbot, J. H.: Acta Crystallogr. *6*, 720 (1953)
217. Iwaizumi, M.: private communication
218. Daul, C., Schläpfer, C. W., von Zelewsky, A.: Structure and Bonding *36*, 129 (1979)
219. Rudin, M.: Ph. D. Thesis ETHZ, to be published
220. Weber, H. P.: unpublished results
221. McConnell, H. M.: J. Chem. Phys. *24*, 764 (1956)
222. Schweiger, A.: unpublished results
223. Fantucci, P., Valenti, V.: J. Am. Chem. Soc. *98*, 3832 (1976)
224. Nishida, Y., Kida, S.: Bull. Chem. Soc. Japan *51*, 143 (1978)
225. Hitchman, M. A.: Inorg. Chem. *16*, 1985 (1977)
226. McGarvey, B. R.: Can. J. Chem. *53*, 2498 (1975)
227. Schläpfer, C. W.: private communication
228. Jörin, E., Graf, F., Schweiger, A., Günthard, Hs. H.: Chem. Phys. Letters *42*, 376 (1976)
229. Bayston, J. H., Looney, F. D., Pilbrow, J. R., Winfield, M. E.: Biochemistry *9*, 2164 (1970)
230. Schrauzer, G. N., Lee, L. P.: J. Am. Chem. Soc. *90*, 6541 (1968)

231. Schweiger, A., Jörin, E., Günthard, Hs. H.: Chem. Phys. Letters *61*, 223 (1979)
232. Jörin, E., Schweiger, A., Günthard, Hs. H.: Chem. Phys. Letters *61*, 228 (1979)
233. Rubinson, K. A.: J. Am. Chem. Soc. *101*, 6105 (1979)
234. Smith, T. D., Pilbrow, J. R.: in: Biological Magnetic Resonance, Vol. 2; Berliner, L. J., Reuben, J., eds. New York: Plenum Press (1980)
235. Orme-Johnson, W. H., Sands, R. H.: in: Iron-Sulfur Proteins, Vol. 2; Lovenberg, W., ed. New York: Academic Press (1973) p. 195
236. Cammack, R., Dickson, D. P. E., Johnson, C. E.: in: Iron-Sulfur Proteins, Vol. 3; Lovenberg, W., ed. New York: Academic Press (1977) p. 283
237. Palmer, G.: in: The Porphyrins, Vol. 4; Dolphin, D., ed. New York: Academic Press (1979)
238. Scholes, C. P.: Proc. Nat. Acad. Sci. US *62*, 428 (1969)
239. Scholes, C. P.: J. Chem. Phys. *52*, 4890 (1970)
240. Scholes, C. P.: in: Multiple Electron Resonance Spectroscopy; Dorio, M. M., Freed, J. H., eds. New York: Plenum Press (1979)
241. Scholes, C. P., Isaacson, R. A., Feher, G.: Biochim. Biophys. Acta *263*, 448 (1972)
242. Mun, S. K., Chang, J. C., Das, T. P.: Biochim. Biophys. Acta *490*, 249 (1977)
243. Feher, G., Isaacson, R. A., Scholes, C. P.: Ann. N. Y. Acad. Sci. *222*, 86 (1973)
244. Heller, P.: Amer. J. Med. *41*, 799 (1966)
245. Mulks, C. F., Scholes, C. P., Dickinson, L. C., Lapidot, A.: J. Am. Chem. Soc. *101*, 1645 (1979)
246. Scholes, C. P., Isaacson, R. A., Yonetani, T., Feher, G.: Biochim. Biophys. Acta *322*, 457 (1973)
247. Scholes, C. P., van Camp, H. L.: Biochim. Biophys. Acta *434*, 290 (1976)
248. Helcké, G. A., Ingram, D. J. E., Slade, E. F.: Proc. R. Soc. London, Ser. B *169*, 275 (1968)
249. Peisach, J., Mims, W. B., Davis, J. L.: J. Biol. Chem. *254*, 12379 (1979)
250. Griffin, B. W., Peterson, J. A.: J. Biol. Chem. *250*, 6445 (1975)
251. Tsai, R., Yu, C. A., Gunsalus, I. C., Peisach, J., Blumberg, W., Orme-Johnson, W. H., Beinert, H.: Proc. Natl. Acad. Sci. USA *66*, 1157 (1970)
252. LoBrutto, R., Scholes, C. P., Wagner, G. C., Gunsalus, I. C., Debrunner, P. G.: J. Am. Chem. Soc. *102*, 1167 (1980)
253. Philson, S. B., Debrunner, P. G., Schmidt, P. G., Gunsalus, I. C.: J. Biol. Chem., in press
254. King, T. E., LoBrutto, R., Scholes, C. P., Wei, Y. H., Yoshida, S.: Presented at the Symposium on Interaction between Iron and Proteins in Oxygen and Electron Transport, Airlie House, Airlie, Virginia, April 13–18 (1980)
255. van Camp, H. L., Scholes, C. P., Mulks, C. F.: J. Am. Chem. Soc. *98*, 4094 (1976)
256. Mallick, M. K., Chang, J. C., Das, T. P.: J. Chem. Phys. *68*, 1462 (1978)
257. Scholes, C. P., Isaacson, R. A., Feher, G., Inubuski, T., Lapidot, A.: in preparation
258. Tsibris, J. C. M., Woody, R. W.: Coord. Chem. Rev. *5*, 417 (1970)
259. Averill, B. A., Orme-Johnson, W. H.: in: Metal Ions in Biology, Vol. 7; Siegel, H., ed. New York: Marcel Dekker (1978) p. 127
260. Cammack, R.: in: Metalloproteins: Structure, Function and Clinical Aspects; Weser, V., ed. Stuttgart: Thieme (1979) p. 162
261. Hall, D. O., Cammack, R., Rao, K. K.: in: Iron in Biochemistry and Medicine; Jacob, A., Worwood, M., eds. London: Academic Press (1974)
262. Palmer, G.: in: The Enzymes, Vol. 12; 3rd. ed., Boyer, P. D., ed. New York: Academic Press (1975)
263. Lovenberg, W., ed.: Iron-Sulfur Proteins, Vols. 1–3, New York: Academic Press (1977)
264. Xavier, A. V., Moura, J. J. G., Moura, I.: Structure and Bonding *43*, 187 (1981)
265. Sands, R. H.: in: Multiple Electron Resonance Spectroscopy. Dorio, M. M., Freed, J. H., eds. New York: Plenum Press (1979)
266. Shetna, Y. I., Wilson, P. W., Hansen, R. E., Beinert, H.: Proc. Nat. Acad. Sci. USA *55*, 1263 (1964)
267. Dunham, W. R., Palmer, G., Sands, R. H., Bearden, A. J.: Biochim. Biophys. Acta *253*, 373 (1971)
268. Fritz, J., Anderson, R., Fee, J., Palmer, G., Sands, R. H., Orme-Johnson, W. H., Beinert, H., Tsibris, J. C., Gunsalus, I. C.: Biochim. Biophys. Acta *253*, 110 (1971)

269. Anderson, R. E., Dunham, W. R., Sands, R. H., Bearden, A. J., Crespi, H. L.: Biochim. Biophys. Acta *408*, 306 (1975)
270. Mukai, K., Kimura, T., Helbert, J., Kevan, L.: Biochim. Biophys. Acta *295*, 49 (1973)
271. Bowman, M., Kevan, L., Mukai, K., Kimura, T.: Biochim. Biophys. Acta *328*, 244 (1973)
272. Reid, J. A.: Ph. D. Thesis, University of Michigan (1976)
273. Anderson, R. E., Anger, G., Peterson, L., Ehrenberg, A., Cammack, R., Hall, D. O., Mullinger, R., Rao, K. K.: Biochim. Biophys. Acta *376*, 63 (1974)
274. Deiss, E.: Ph. D. Thesis ETHZ Nr. 6659
275. Hausser, K. H.: Z. Naturforsch. *16a*, 1190 (1961)
276. Anderson, S. E., Drago, R. S.: J. Am. Chem. Soc. *91*, 3656 (1969)
277. Förster, E., Albrecht, G., Dürselen, W., Kurras, E.: J. Organomet. Chem. *19*, 215 (1969)
278. Seiler, P., Dunitz, J. D.: Acta Cryst. *B35*, 1068 (1979)
279. Seiler, P., Dunitz, J. D.: Acta Cryst. *B35*, 2020 (1979)
280. Campbell, A. J., Fyfe, C. A., Harold-Smith, D., Jeffrey, K. R.: Mol. Cryst. Liquid Cryst. *36*, 1 (1976)
281. Wolf, R., Schweiger, A., Günthard, Hs. H.: in preparation
282. Labauze, G., Raynor, J. B., Samuel, E.: J. C. S. Dalton, 2425 (1980)
283. Samuel, E., Labauze, G., Vivien, D.: J. C. S. Dalton 956 (1979)
284. Rettig, M. F., Stout, C. D., Klug, A., Farnham, P.: J. Am. Chem. Soc. *92*, 5100 (1970)
285. Prins, R., Reinders, F. J.: Chem. Phys. Letters *3*, 45 (1969)
286. Kokin, R. P.: Mag. Res. Rev. *5*, 75 (1979)
287. Chasteen, N. D.: Coord. Chem. Rev. *22*, 1 (1977)
288. van Willigen, H., Mulks, C. F., Bouhaouss, A., Ferhat, M., Roufosse, A. H.: J. Am. Chem. Soc. *102*, 4846 (1980)
289. van Willigen, H.: Chem. Phys. Letters *65*, 490 (1979)
290. van Willigen, H.: J. Mag. Res. *39*, 37 (1980)
291. van Willigen, H.: privat communication
292. Fujimoto, M., Dressel, L. A.: Ferroelectrics *8*, 611 (1974)
293. Möhl, W., Winscom, C. J., Plato, M., Möbius, K., Lubitz, W., van Willigen, H.: Bull. Mag. Res. *2*, 219 (1981)
294. Winscom, C. J.: Mol. Phys. *28*, 1579 (1974)

Subject Index

Author-Index Volumes 1–51

Inorganic Chemistry Concepts

Editors: C.K. Jørgensen, M.F. Lappert, S.J. Lippard, J.L. Margrave, K. Niedenzu, H. Nöth, R.W. Parry, H. Yamatera

Volume 6
D.L. Kepert

Inorganic Stereochemistry

1981. 206 figures, 45 tables. Approx. 250 pages
ISBN 3-540-10716-9

Contents: Introduction. - Polyhedra. - Four-Coordinate Compounds. - Five-Coordinate Compounds Containing only Unidentate Ligands. - Five-Coordinate Compounds Containing Chelate Groups. - Six-Coordinate Compounds Containing only Unidentate Ligands. - Six-Coordinate Compounds [M(Bidentate)$_2$(Unidentate)$_2$]. - Six-Coordinate Compounds [M(Bidentate)$_3$]. - Six-Coordinate Compounds Containing Tridentate Ligands. - Seven-Coordinate Compounds Containing only Unidentate Ligands. - Seven-Coordinate Compounds Containing Chelate Groups. - Eight-Coordinate Compounds Containing only Unidentate Ligands. - Eight-Coordinate Compounds Containing Chelate Groups. - Nine-Coordinate Compounds. - Ten-Coordinate Compounds. - Twelve-Coordinate Compounds. - References. - Subject Index.

Volume 5
T. Tominaga, E. Tachikawa

Modern Hot-Atom Chemistry and Its Applications

1981. 57 figures. VIII, 155 pages
ISBN 3-540-10715-0

Contents: Introduction. - Experimental Techniques: Production of Energetic Atoms. Radiochemical Separation Techniques. Special Physical Techniques. - Characteristics of Hot Atom Reactions: Gas Phase Hot Atom Reactions. Liquid Phase Hot Atom Reactions. Solid Phase Hot Atom Reactions. - Applications of Hot Atom Chemistry and Related Topics: Applications in Inorganic, Analytical and Geochemistry. Applications in Physical Chemistry. Applications in Biochemistry and Nuclear Medicine. Hot Atom Chemistry in Energy-Related Research. Current Topics Related to Hot Atom Chemistry and Future Scope. - Subject Index.

Volume 4
Y. Saito

Inorganic Molecular Dissymmetry

1979. 107 figures, 28 tables. IX, 167 pages
ISBN 3-540-09176-9

Contents: Introduction. - X-Ray Difraction. - Conformational Analysis. - Structure and Isomerism of Optically Active Complexes. - Electron-Density Distribution in Transition Metal Complexes. - Circular Dichroism. - References. - Subject Index.

Volume 3
P. Gütlich, R. Link, A. Trautwein

Mössbauer Spectroscopy and Transition Metal Chemistry

1978. 160 figures, 19 tables, 1 folding plate. X, 280 pages
ISBN 3-540-08671-4

Contents: Basic Physical Concepts. - Hyperfine Interactions. - Experimental. - Mathematical Evaluation of Mössbauer Spectra. - Interpretation of Mössbauer Parameters of Iron Compounds. - Mössbauer-Active Transition Metals Other than Iron. - Some Special Applications.

Volume 2
R.L. Carlin, A.J. van Duyneveldt

Magnetic Properties of Transition Metal Compounds

1977. 149 figures, 7 tables. XV, 264 pages
ISBN 3-540-08584-X

Contents: Paramagnetism: The Curie Law. - Thermodynamics and Relaxation. - Paramagnetism: Zero-Field Splittings. - Dimers and Clusters. - Long-Range Order. - Short-Range Order. - Special Topics: Spin-Flop, Metamagnetism, Ferrimagnetism and Canting. - Selected Examples.

Volume 1
R. Reisfeld, C.K. Jørgensen

Lasers and Excited States of Rare Earths

1977. 9 figures, 26 tables. VIII, 226 pages
ISBN 3-540-08324-3

Contents: Analogies and Differences Between Monatomic Entities and Condensed Matter. - Rare-Earth Lasers. - Chemical Bonding and Lanthanide Spectra. - Energy Transfer. - Applications and Suggestions.

Springer-Verlag
Berlin
Heidelberg
New York

DATE DUE

GAYLORD			PRINTED IN U.S.A.